THE IMPACT OF ACADEMIC RESEARCH ON INDUSTRIAL PERFORMANCE

NATIONAL ACADEMY OF ENGINEERING
OF THE NATIONAL ACADEMIES

THE NATIONAL ACADEMIES PRESS
Washington, D.C.
www.nap.edu

THE NATIONAL ACADEMIES PRESS • 500 Fifth Street, N.W. • Washington, D.C. 20001

This study was supported by Grant No. 97-3-4 between the National Academy of Sciences and the Alfred P. Sloan Foundation and by the National Academy of Engineering Fund. Any opinions, findings, conclusions, or recommendations expressed in this publication are those of the author(s) and do not necessarily reflect the views of the organizations or agencies that provided support for the project.

International Standard Book Number 0-309-08973-5 (Book)
International Standard Book Number 0-309-52602-7 (PDF)

Copies of this report are available from the National Academies Press, 500 Fifth Street, N.W., Lockbox 285, Washington, DC 20055; (800) 624-6242 or (202) 334-3313 (in the Washington metropolitan area); online at *http://www.nap.edu.*

Printed in the United States of America

THE NATIONAL ACADEMIES

Advisers to the Nation on Science, Engineering, and Medicine

The **National Academy of Sciences** is a private, nonprofit, self-perpetuating society of distinguished scholars engaged in scientific and engineering research, dedicated to the furtherance of science and technology and to their use for the general welfare. Upon the authority of the charter granted to it by the Congress in 1863, the Academy has a mandate that requires it to advise the federal government on scientific and technical matters. Dr. Bruce M. Alberts is president of the National Academy of Sciences.

The **National Academy of Engineering** was established in 1964, under the charter of the National Academy of Sciences, as a parallel organization of outstanding engineers. It is autonomous in its administration and in the selection of its members, sharing with the National Academy of Sciences the responsibility for advising the federal government. The National Academy of Engineering also sponsors engineering programs aimed at meeting national needs, encourages education and research, and recognizes the superior achievements of engineers. Dr. Wm. A. Wulf is president of the National Academy of Engineering.

The **Institute of Medicine** was established in 1970 by the National Academy of Sciences to secure the services of eminent members of appropriate professions in the examination of policy matters pertaining to the health of the public. The Institute acts under the responsibility given to the National Academy of Sciences by its congressional charter to be an adviser to the federal government and, upon its own initiative, to identify issues of medical care, research, and education. Dr. Harvey V. Fineberg is president of the Institute of Medicine.

The **National Research Council** was organized by the National Academy of Sciences in 1916 to associate the broad community of science and technology with the Academy's purposes of furthering knowledge and advising the federal government. Functioning in accordance with general policies determined by the Academy, the Council has become the principal operating agency of both the National Academy of Sciences and the National Academy of Engineering in providing services to the government, the public, and the scientific and engineering communities. The Council is administered jointly by both Academies and the Institute of Medicine. Dr. Bruce M. Alberts and Dr. Wm. A. Wulf are chair and vice chair, respectively, of the National Research Council.

www.national-academies.org

COMMITTEE ON THE IMPACT OF ACADEMIC RESEARCH ON INDUSTRIAL PERFORMANCE

JEROME H. GROSSMAN (IOM), *chair*, Lion Gate Management Corporation and Harvard University, Cambridge, Massachusetts
LILLIAN C. BORRONE (NAE), Port Authority of New York and New Jersey (retired), New York, New York
COLIN CROOK, Citicorp (retired), New York, New York
DONALD M. ENGELMAN (NAS), Yale University, New Haven, Connecticut
DAVID J. FARBER, University of Pennsylvania, Philadelphia
ANNETINE C. GELIJNS, Columbia Medical Center, New York, New York
DANIEL GREGORY, Greylock Management Corporation, Boston, Massachusetts
GEORGE H. HEILMEIER (NAE), Telcordia Technologies, Inc. (retired), Morristown, New Jersey
ADAM B. JAFFE, Brandeis University, Waltham, Massachusetts
JACK L. KERREBROCK (NAE), Massachusetts Institute of Technology (retired), Cambridge
KENT KRESA (NAE), Northrop Grumman Corporation, Los Angeles, California
H. DONALD RATLIFF (NAE), Georgia Institute of Technology, Atlanta
J. DAVID ROESSNER, Georgia Institute of Technology, Atlanta
ROBERT F. SPROULL (NAE), Sun Microsystems, Inc., Burlington, Massachusetts
MORRIS TANENBAUM (NAE), AT&T Corporation (retired), Short Hills, New Jersey

PANEL ON THE NETWORK SYSTEMS AND COMMUNICATIONS INDUSTRY

ROBERT F. SPROULL (NAE), *chair,* Sun Microsystems, Inc., Burlington, Massachusetts
ALFRED V. AHO (NAE), Columbia University, New York, New York
JOHN M. CIOFFI (NAE), Stanford University, Stanford, California
DAVID J. FARBER, University of Pennsylvania, Philadelphia
GEORGE H. HEILMEIER (NAE), Telcordia Technologies, Inc. (retired), Morristown, New Jersey
STEPHEN WOLFF, Cisco Systems, Inc., Washington, D.C.

PANEL ON THE MEDICAL DEVICES AND EQUIPMENT INDUSTRY

ANNETINE C. GELIJNS, *chair,* Columbia Medical Center, New York, New York
PAUL CITRON (NAE), Medtronic, Inc., Minneapolis, Minnesota
DONALD M. ENGELMAN (NAS), Yale University, New Haven, Connecticut
JEROME H. GROSSMAN (IOM), Lion Gate Management Corporation and Harvard University, Cambridge, Massachusetts
HOMER PIEN, SRU Biosytems, Woburn, Massachusetts
FREDERICK W. TELLING, Pfizer Inc., New York, New York

PANEL ON THE AEROSPACE INDUSTRY

JACK L. KERREBROCK (NAE), *chair,* Massachusetts Institute of Technology, Cambridge
ANTONIO L. ELIAS (NAE), Orbital Sciences Corporation, Dulles, Virginia
KENNETH C. HALL, Duke University, Durham, North Carolina
ROBERT J. HERMANN (NAE), Global Technology Partners, LLC, Hartford, Connecticut
KENT KRESA (NAE), Northrop Grumman Corporation, Los Angeles, California

PANEL ON THE TRANSPORTATION, DISTRIBUTION, AND LOGISTICS SERVICES INDUSTRY

H. DONALD RATLIFF (NAE), *chair,* Georgia Institute of Technology, Atlanta
CYNTHIA BARNHART, Massachusetts Institute of Technology, Cambridge
ROBERT E. BIXBY (NAE), ILOG, Inc., Houston, Texas
LILLIAN C. BORRONE (NAE), Port Authority of New York and New Jersey (retired), New York, New York
A. RAY CHAMBERLAIN, Parsons Brinckerhoff, Inc., Denver, Colorado
ADAM B. JAFFE, Brandeis University, Waltham, Massachusetts

PANEL ON THE FINANCIAL SERVICES INDUSTRY

COLIN CROOK, *chair,* Citicorp (retired), New York, New York
CAMILO C. GOMEZ, private consultant, Santa Fe, New Mexico
ADAM B. JAFFE, Brandeis University, Waltham, Massachusetts
BLAKE D. LeBARON, Brandeis University, Waltham, Massachusetts
JOHN M. MULVEY, Princeton University, Princeton, New Jersey
MORRIS TANENBAUM (NAE), AT&T Corporation (retired), Short Hills, New Jersey

NAE PROGRAM OFFICE STAFF

PROCTOR P. REID, Study Director
DIANE ALBERT, J. Herbert Hollomon Fellow (until January 1999)
CAROL R. ARENBERG, Managing Editor
THOMAS C. MAHONEY, Consultant
ROBERT P. MORGAN, NAE Fellow and Senior Analyst (through January 1999), and Consultant (after January 1999)
PENELOPE J. GIBBS, Program Associate
NATHAN KAHL, Senior Project Assistant

Preface

American universities are integral to the success of American industry. The core mission of the university, educating much of the American workforce, is essential to productivity growth, innovation, technological progress, and virtually every other national economic and societal objective. Academic research—in physical sciences, social sciences, computer sciences, and engineering—provides a constant flow of ideas, analyses, and breakthroughs that vitalize industry. The strengths of academic research—principally the resources to focus on long-term, fundamental, risky goals and to mount broad collaborative projects—complement the applied research and development (R&D) performed by industry. Universities are a source not only of scientific and technological ideas that lead to new products and processes, but also social and political insights that strengthen the nation's ability to adapt to new technologies and, therefore, to embrace continued innovation. As industries have become more dependent on innovation, new skills, and technological prowess, academic contributions have become increasingly critical to economic success.

To assess and document the contributions of academic research to industry, the National Academy of Engineering initiated a study of the impact of academic research on five diverse industries: network systems and communications; financial services; medical devices and equipment; transportation, distribution, and logistics services; and aerospace. All five industries are important in terms of sales and employment, technological intensity, and expected growth rates; each provides a distinct example of current and historic patterns of interaction with academia. The study is based on the opinions and judgments of a 15-member committee of experts from industry and universities, supported by five panels (one for each industry) that reported to the study committee. The committee's deliberations were informed by surveys of industry executives and leading academics, workshop discussions, and reviews of relevant publications.

These five industries illustrate the wide range of contributions of academic research to industrial performance: graduates trained in modern research techniques; fundamental concepts and key ideas emerging from basic and applied research; and the development of tools, prototypes, and marketable products, processes, and services. The impact of academic research derives from many disciplines, including the natural sciences and engineering as well as the social and behavioral sciences, with the value of research in one field often heavily dependent on advances in complementary fields. Contributions also emerge from the broad base of knowledge resident at universities, which provide environments where ideas developed in one context often flourish in very different contexts. This broad knowledge base often results in contributions in areas essential to successful innovation but beyond any specific technology or field; examples include industry regulation and deregulation and the development of industry standards.

The study identifies major cross-cutting challenges for university-industry research collaboration, including a growing imbalance in federal funding for academic research, an underdeveloped interface between research universities and services industries (roughly 80 percent of the U.S. economy), and reconciling traditional university missions of teaching, research, and service with the increased emphasis on the management of intellectual property. The committee recommends actions to meet these challenges for government, universities, and industry.

On behalf of the National Academy of Engineering, I want to thank the study chairman, Jerome H. Grossman, the chairs of the industry panels, Colin Crook, Annetine C. Gelijns, Jack L. Kerrebrock, H. Donald Ratliff, and Robert Sproull, and other members of the study committee and the five industry panels (named on pp. iv-vi) for their considerable efforts on this project. I also want to thank Proctor P. Reid, the study director, who managed the project and helped the committee members reach consensus. Thomas C. Mahoney, consultant to the committee and lead staff support to the aerospace industry panel, and Robert P. Morgan, former NAE Fellow and lead staff support to the panels on the medical devices and equipment industry and the network systems and communications industry were extremely helpful throughout the project, particularly during the closing phase of the study. Diane Albert, former NAE J. Herbert Hollomon Fellow, provided lead staff support to the panel on transportation, distribution, and logistics services throughout the fact-finding phase. Penelope Gibbs and Nathan Kahl from the NAE Program Office provided critical administrative and logistical support. Carol Arenberg, NAE managing editor, was instrumental in preparing the report for publication.

I want to extend the committee's thanks to everyone from government, industry, and academia who contributed to the project. In particular, I want to express our appreciation to the participants in the fact-finding workshops, roundtables, and e-mail surveys during the initial stages of the project (see addenda to chapters 2 through 6) and to those who briefed the committee.

Finally, I would like to express my appreciation to the Alfred P. Sloan Foundation for its generous support of this project.

Wm. A. Wulf
President
National Academy of Engineering

Acknowledgments

This report has been reviewed in draft form by individuals chosen for their diverse perspectives and technical expertise, in accordance with procedures approved by the NRC's Report Review Committee. The purpose of this independent review is to provide candid and critical comments that will assist the institution in making its published report as sound as possible and to ensure that the report meets institutional standards for objectivity, evidence, and responsiveness to the study charge. The review comments and draft manuscript remain confidential to protect the integrity of the deliberative process. We wish to thank the following individuals for their review of this report: Chris Caplice, SABRE Group; Irwin Feller, Pennsylvania State University; James Flanagan, Rutgers University; James F. Gibbons, Stanford University; Cliff Goodman, The Lewin Group; Patrick T. Harker, University of Pennsylvania; David Heebner, Heebner Associates; Kenneth H. Keller, University of Minnesota; Ed Lazowska, University of Washington; Robert W. Lucky, Telcordia Technologies, Inc.; Hans Mark, University of Texas at Austin; Richard Nelson, Columbia University; George L. Nemhauser, Georgia Institute of Technology; John Parrish, Harvard Medical School and Massachusetts General Hospital; Sholom Rosen, Citibank (retired); Chester S. Spatt, Carnegie Mellon University; Donald E. Strickland, Southern Illinois University; Marie C. Thursby, Georgia Institute of Technology; Chelsea C. White III, Georgia Institute of Technology; David C. Wisler, GE Aircraft Engines; Ben T. Zinn, Georgia Institute of Technology

Although the reviewers listed above have provided many constructive comments and suggestions, they were not asked to endorse the conclusions or recommendations nor did they see the final draft of the report before its release. The review of this report was overseen by NAE member Elsa Garmire, Dartmouth

College appointed by the National Academy of Engineering, who was responsible for making certain that an independent examination of this report was carried out in accordance with institutional procedures and that all review comments were carefully considered. Responsibility for the final content of this report rests entirely with the authoring committee and the institution.

Contents

Executive Summary

Universities have a long history of contributing to industry and the economy in a variety of ways. Yet only recently has attention been focused systematically on the nature and extent of these contributions to specific industries. With support from the Alfred P. Sloan Foundation, the National Academy of Engineering initiated a study in 1998 to document and assess the contributions of academic research to the performance of five industry sectors: network systems and communications; medical devices and equipment; aerospace; transportation, distribution, and logistics services; and financial services. A 15-member committee of industry and university experts conducted the study, with integral support from five panels (one for each industry).

Since this study began, there has been growing recognition of the importance of universities and academic research to industrial innovation and performance. There has also been greater recognition of how the role of academic research has changed over time, especially over the last 25 years as many universities have become more directly involved in the commercialization of the fruits of their research. This study documents the contributions of university research to five very different industry sectors and provides a qualitative assessment of the impact of that research. The committee's conclusions cut across all five sectors; the recommendations propose steps that should be taken to meet a variety of challenges in academic research and industry-university interactions.

Few data are available to support a quantitative assessment of the impact of academic research on the performance of specific industries. Quantitative measures of the output of academic research seldom go beyond counts of research papers, patents, and royalty income, none of which directly correlates with the impact on industry. Industrial performance can be measured by shareholder value, employment growth, market share, technical advances, and other factors, all of which tend to be cyclical and company specific. Therefore, the committee relied on informed opinion and expert judgments, supplemented by reviews of the literature, e-mail surveys, workshop discussions, and panel deliberations, to make qualitative assessments of the impact of academic research.

All five industry sectors examined in this study are important in terms of sales and employment, technological intensity, and expected growth rates. Two sectors, medical devices and equipment and network systems and communications, have a recent history of extensive collaboration with academic researchers. The other three, aerospace, financial services, and transportation, distribution, and logistics services, have had more limited, less systematic interactions with the academic research enterprise. All five have demonstrated significant capacities for innovation over the past decade. However, the financial services industry and the transportation, distribution, and logistics services industry, the two predominantly service industries, have only recently begun to integrate research and development (R&D) into their way of doing business. Ultimately, the committee concluded that academic research has had a significant impact on performance in the network systems and communications, medical devices and equipment, and financial services industries. The impact on performance in the aerospace and the transportation, distribution, and logistics services industries has been moderate.

Academic research has made substantial contributions to all five industries, ranging from graduates at all levels trained in modern research techniques to fundamental concepts and key ideas based on basic and applied research to the development of tools, prototypes, and marketable products, processes, and services. The disciplinary sources of research contributions span the fields of engineering, the natural sciences, and the social and behavioral sciences, with advances in one field often combining and building on developments in other fields. Pathways linking academic research and industries include: direct hires of students, graduates, and faculty; temporary exchanges of researchers; faculty consultancies; joint research involving industry and academic scientists and engineers; industry-sponsored research contracts and grants; a variety of institutional mechanisms at universities (e.g., research centers, consortia, industrial liaison programs); technology licensing; start-up companies; publications; conferences; and short courses.

Each industry studied illustrates a distinct pattern of collaboration with universities and has developed different mechanisms for taking advantage of academic contributions. For example, network systems has a history of drawing upon academic research for fundamental innovations, as well as using universities as test beds for new networking concepts that have provided the underpinnings of the Internet, the World Wide Web, and e-commerce. Figure ES-1 shows the complex pattern of academic and industrial research over many years in this industry. The medical devices and equipment industry has also looked to universities for fundamental multidisciplinary research in the physical sciences and engineering and the unique capabilities of academic medical centers for researching, developing, testing, and improving devices, as well as conducting the clinical trials necessary to obtaining regulatory approval, all in an atmosphere of close industry-university collaboration. In financial services, contributions of academic research in economics, engineering, and mathematics have been important to the development of new financial models and instruments, in spite of the industry's lack of a well developed R&D infrastructure.

This study confirms that graduates trained in research are a major component of academia's contribution to industrial performance. U.S. universities are industries' primary source of people with research training and experience—undergraduates, graduate students, postdoctoral researchers, and faculty. Research-trained students and graduates at all degree levels play a critical role in the development, transfer, diffusion, and application of new knowledge and technology both within and between industry and academia.

A second consistent message that emerges from a study of these five industries is that basic, long-term research performed by universities across a wide range of science and engineering disciplines has made significant contributions to industry over time. For instance, portfolio theory, linear programming, derivative pricing theory, and prospect theory, all of academic origin, have laid the foundation for whole new families of financial products and services. Academic contributions to linear and integer programming and to queue theory are the building blocks of the information-management and decision-support technologies at the heart of the integrated logistics revolution. Medical devices, such as magnetic resonance imaging machines and pacemakers, are based on the contributions of fundamental research arising from multiple disciplines in the natural sciences and engineering.

Universities perform roughly half of all basic research in the United States, most of it funded by the federal government. Over time, basic research results build on one another and intermingle with results in other fields, often through the free exchange of people and ideas that universities facilitate. Typically pursued with no specific commercial application in mind, basic research has provided the technological underpinnings for commercial innovation. Relevant research findings have come from multiple disciplines, indicating that a portfolio of research investment in many different fields is essential to continued progress. In some cases, the original academic source of a basic understanding of a process or technology is forgotten by the time the knowledge is put to use; in other cases, Nobel prizes are awarded years later for fundamental breakthroughs. Individual academic researchers may or may not receive the credit they deserve for the constant flow of commercial innovations, but there can be no doubt that basic research and the publication of its results have made a unique and essential contribution to industrial performance.

Contributions from applied academic research are also very important to industry. Today, academic applied researchers and their academic research

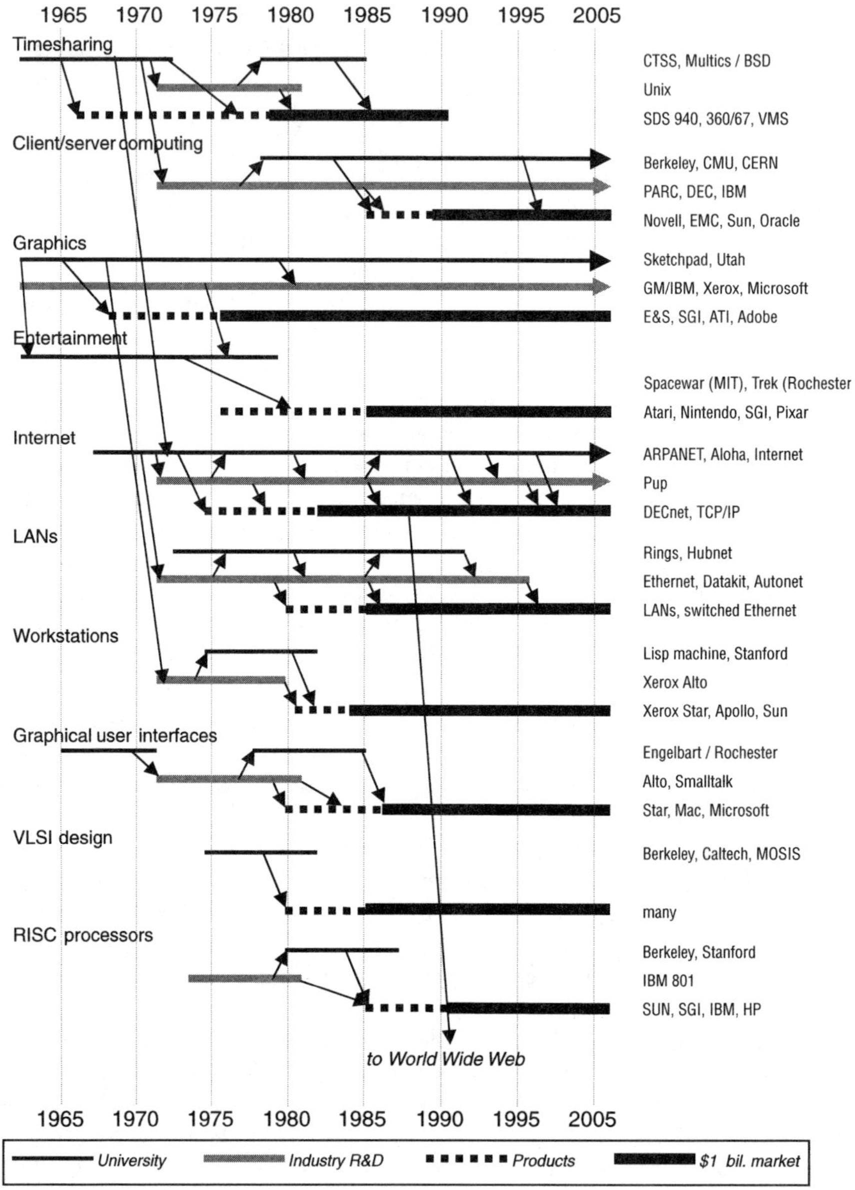

FIGURE ES-1 Examples of academic government-sponsored (and some industry-sponsored) IT research and development in the creation of commercial products and industries. Source: NRC, 2003.

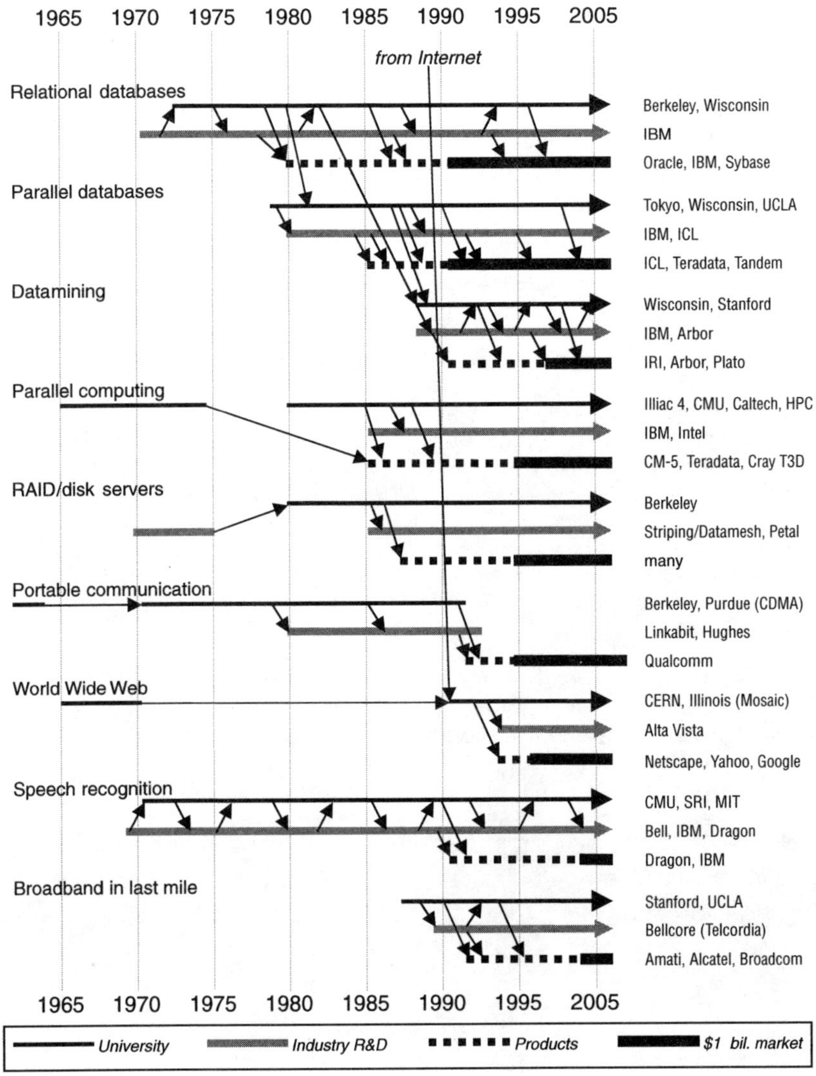
1965 1970 1975 1980 1985 1990 1995 2005
from Internet
Relational databases
Berkeley, Wisconsin
IBM
Oracle, IBM, Sybase
Parallel databases
Tokyo, Wisconsin, UCLA
IBM, ICL
ICL, Teradata, Tandem
Datamining
Wisconsin, Stanford
IBM, Arbor
IRI, Arbor, Plato
Parallel computing
Illiac 4, CMU, Caltech, HPC
IBM, Intel
CM-5, Teradata, Cray T3D
RAID/disk servers
Berkeley
Striping/Datamesh, Petal
many
Portable communication
Berkeley, Purdue (CDMA)
Linkabit, Hughes
Qualcomm
World Wide Web
CERN, Illinois (Mosaic)
Alta Vista
Netscape, Yahoo, Google
Speech recognition
CMU, SRI, MIT
Bell, IBM, Dragon
Dragon, IBM
Broadband in last mile
Stanford, UCLA
Bellcore (Telcordia)
Amati, Alcatel, Broadcom
1965 1970 1975 1980 1985 1990 1995 2005
University
Industry R&D
Products
$1 bil. market

infrastructure are directly involved in the development of industrial tools, prototypes, products, and production processes, as well as the delivery of products and services. Box ES-1 shows examples of specific applied academic research contributions to each of the five industries. Sometimes academic applied research yields cumulative, incremental advances over time that prove to be of major importance to industry. Individual companies have also greatly benefited from university-based research to solve discrete practical problems related to

BOX ES-1
Contributions of Applied Academic Research

Network Systems and Communications. Contributions include packet switching and the Internet TCP/IP protocol, both key elements in the development of the Internet. The Mosaic web browser interface was an important step in the rapid evolution of the worldwide web. University researchers and other university personnel have contributed in significant ways to routers, the development of ATM switches, digital subscriber line (DSL) technology, third-generation wireless transmission, computer graphics, search engines, traffic management, stable broadcast networking, the evolution of new networks, and the development of standards.

Medical Devices and Equipment. The development of a wide range of therapeutic and diagnostic devices has resulted from the involvement of academic researchers and academic medical centers in R&D, prototype testing, evaluation, and clinical trials. Devices and equipment include magnetic resonance imaging equipment; whole-body CAT scanners; flexible endoscopy; lasers for a broad range of medical applications, from gastrointestinal surgery to eye surgery; cardiac assist devices; organ and joint replacements; ultrasound and minimally invasive surgical techniques; and advances in tissue engineering.

Aerospace. Contributions to the development of tools, include advanced nonintrusive instrumentation, flow visualization techniques, and computational fluid dynamics. Contributions to specific technologies have been made in the areas of heat transfer, combustion cooling, and aeromechanics; low-Reynolds-number airfoil design; Internet by satellite, including protocols and computational tools for data integration; and folding-wing design for small unmanned aerial vehicles.

Transportation, Distribution, and Logistics Services. Contributions include optimization modeling for shippers, software applications/decision support systems for routing, production scheduling, logistics, and distribution management. Academic spin-off companies have commercialized much of this software.

Financial Services. Contributions include new financial instruments, including index funds and derivatives; financial information and research tools, including risk/credit metrics and financial risk management software; models for pricing derivatives and securities; and advances in cryptography for specific financial services.

their businesses. University-based research centers, with industrial participation, have become another important venue for applied research as well as more directed basic research of value to industry. Applied research through multidisciplinary collaboration among science, engineering, and/or medical faculties is a unique strength of academia. Most of the funding for applied research at universities comes from federal agencies that want specific problems solved related to their missions. Industry funds some academic applied research directly via research contracts and also indirectly through support of university-based research centers. Industry funding of academic research increased throughout the 1990s, but industry still provides only a small proportion of total funds for academic research.

The natural sciences and engineering disciplines are not the only fields that make important research contributions to industrial innovation and performance. The contributions of the social and behavioral sciences to industry have been greatly undervalued. Researchers in business, economics, psychology, and many other fields have made valuable contributions to progress in the industries studied. For example, in network systems and communications, social and behavioral scientists have provided the knowledge base for the formulation of regulatory policy and have generated useful knowledge and insights into group dynamics and decision making, the diffusion of new technologies, the nature and value of network externalities, and the relationship between organizational characteristics and information dissemination and sharing.

CHALLENGES AND OPPORTUNITIES

The industry studies revealed six major, crosscutting challenges and opportunities for university-industry research that warrant careful attention by universities, industry, and government.

First, there is a growing imbalance in federal R&D funding. Current investments in life sciences far outpace investments in the complementary disciplines of physical sciences, engineering, and the social and behavioral sciences. The industry studies show time and again that the value of research results in one field often depends heavily on advances in complementary fields. In addition, universities must maintain a mix of basic and applied research to sustain their role as repositories of expertise and resources in many disciplines. Federal funding is now virtually the only source of support for basic research, which makes the effective management of federal research programs of paramount importance. Program managers in federal agencies must work with academic and industrial researchers to develop research agendas that might lead to major new insights. The challenge is not just to maintain a balance between basic and applied research, but also to ensure that the basic research portfolio is sufficiently diverse to stimulate innovative thinking by academic researchers in many fields.

Second, the industry studies underscore the need for industries and universities to continue exploring mechanisms and pathways for bringing the benefits of academic research to industry, keeping in mind that what works well in one industry may not work well in another. Both partners should experiment with new approaches. University-industry linkages must be adaptable, and universities should be on the lookout for opportunities to link up with new industries and explore leading-edge industrial research activities and challenges. Cross-sectoral movement and interaction between individual academic and industrial researchers are essential to promoting the effective two-way exchange of knowledge and technology.

Third, the studies suggest that services industries represent a significant source of opportunity for university-industry interaction. Services account for more than 80 percent of the U.S. gross domestic product, employ a large and growing share of the science and engineering workforce, and are the primary users of information technology. In most manufacturing industries, service functions (such as logistics, distribution, and customer service) are now leading areas of competitive advantage. Innovation and increased productivity in the services infrastructure (e.g., finance, transportation, communication, health care) have an enormous impact on productivity and performance in all other segments of the economy. Nevertheless, the academic research enterprise has not focused on or been organized to meet the needs of service businesses. Major challenges to services industries that could be taken up by universities include: (1) the adaptation and application of systems and industrial engineering concepts, methodologies, and quality-control processes to service functions and businesses; (2) the integration of technological research and social science, management, and policy research; and the (3) the education and training of engineering and science graduates prepared to deal with management, policy, and social issues.

Fourth, in recent decades, universities have increasingly emphasized technology transfer and the generation of income from research activities by patenting and licensing research results and the creation of technology transfer offices. The increased attention to the management of intellectual property has had many positive consequences for both industry and academia, including providing incentives for invention and innovation and the dissemination and commercialization of new technologies. However, many questions remain regarding the net returns on investments in technology transfer, as well as their impact on the core research and educational missions of universities. These are questions that academic researchers are well equipped to address.

Fifth, regulation and regulatory changes continue to have a profound influence on industry receptivity to contributions by academic research. In turn, academic research continues to shape the regulatory environment of many industries, including most of those in this study. Research universities are well equipped to provide interdisciplinary expertise that can inform ongoing regulatory debates in these and other industries. Moreover, the influx of academically trained scientists and engineers into financial, health, communication, transportation, and other regulatory bodies has strengthened these agencies' ability to draw effectively on research advances across a spectrum of disciplines relevant to their tasks. The rapid pace of technological change in many sectors and its effect on the structure, competitive dynamics, and economic and societal impacts of many industries underscores the importance of sustaining, as well as deepening, the current productive relationship between academic researchers and regulators.

Finally, information technologies are critical to the performance of all industries and will continue to be so in the future. Industry's need for the continued development, diffusion, and effective application of advanced information technologies presents major opportunities for academic research in mathematics, computer sciences, physical sciences, life sciences, multiple engineering disciplines, and social and behavioral sciences.

CONCLUSION

The committee's review of these five very different industries shows that academic research has clearly provided benefits to industry and has had a positive, long-term impact on industrial performance. However, it is difficult to identify specific mechanisms by which this impact can be maximized. The research competencies, the ability to interface with industry, the quality of infrastructure, and many, many other characteristics vary greatly from university to university. In addition, industries in the abstract do not interface with universities; only individual companies do. And companies in a given industry also vary in their ability to manage interfaces with universities, in their expectations of what academic researchers can provide, in the complexity of their research problems, and in their time horizons. And all of these vary over time. When this study began, high-technology industries, such as network systems and communications, were booming, attracting academic researchers and potential graduate students to well funded industrial laboratories, growing operations, and a plethora of start-up companies. As the study comes to a close, this same industry is experiencing decreases in sales, stock prices, investments, research funding, and employment. Thus, both the unique characteristics of individual institutions and the changes brought about by economic cycles must be kept in mind in assessing the impact of academic research.

Ultimately, the study of these five industries underscores the core strengths of the academic research enterprise. Because universities are venues for a greater range of ideas and disciplinary perspectives than any other institutions in the U.S. innovation system, they have vast potential for multidisciplinary research. In addition, universities are the only places where advanced research and education are integrated on a large scale. The constant flow of new students through universities continuously revitalizes the academic research enterprise, challenging the assumptions of faculty and bringing fresh perspectives to research. Indeed, the rich and varied interaction between university research faculty and students and companies in many industries exposes the former to industrial challenges that often serve as stimuli to basic and applied research in academia. These core strengths augur well for the future of academic research and its continuing contributions to industry performance.

GENERAL RECOMMENDATIONS

The general recommendations of this study call for actions that could enhance the contributions of university research to industrial growth and performance. Recommendations for individual industries can be found in the body of the report. The general recommendations address crosscutting challenges and opportunities that apply to more than one industry.

General Recommendation 1. Because the contributions of academic research are diverse and often indirect, a broad and balanced portfolio of academic research should be maintained. Recent trends in federal funding indicate that funding levels for research in the physical sciences, engineering, and the social and behavioral sciences should be increased.

- Congress and the administration should restore the balance in federal funding of academic research by increasing support for research in the physical sciences, engineering, and the social and behavioral sciences to complement and leverage the results of recent heavy investments in the life sciences and medical sciences.
- Federal funding of academic research should continue to emphasize long-term basic research, as well as applied research (typically funded by mission agencies). Multidisciplinary research should be encouraged through support of project-specific research teams and other institutionalized mechanisms, such as engineering research centers and other university-industry research centers.

General Recommendation 2. Industries and universities should continue to explore mechanisms and pathways for bringing the benefits of academic research to industry, keeping in mind that what works well in one industry may not work well in another. Both partners should experiment with new approaches. University-industry research linkages should be adaptable, and universities should be on the lookout for opportunities to link up with new industries and explore leading-edge industrial research activities and challenges.

Given the importance of personal relationships among academic and industrial researchers for productive collaboration and knowledge transfer, universities and industry should foster interactions between university- and industry-based scientists and engineers in the following ways:

- A major program of fellowships should be established to attract and support graduate students in science and engineering.
- Sabbatical programs should be established and/or expanded to encourage academic and industrial researchers to spend time in each other's home research settings.
- More balanced participation by academic researchers and their industry counterparts in major conferences on specific sectors, technical systems, and disciplines should be encouraged.
- New ways of supporting personal interactions across academia-industry boundaries, including using technology to support collaboration, should be explored.
- University-industry research centers should be structured to facilitate close interaction between academic and industrial researchers.
- Academic reward structures, such as promotion and tenure criteria, should be reviewed and modified (as necessary) to encourage and reward researchers who attract research support from industry and/or address significant research questions of direct importance to industry.
- Intellectual property rights policies and practices that facilitate productive research collaboration with industry should be promulgated at universities.

General Recommendation 3. The ability of academic researchers to contribute to services industries and the receptivity of leaders in the services industries to the potential contributions of academic research must both be improved. The following steps would have immediate benefits:

- Academic research contributions and capabilities relevant to each industry should be documented and promoted in the targeted communities to educate senior managers about how academic research might improve company performance in the marketplace.
- Common legal frameworks acceptable to industry and academia should be established detailing the terms of confidentiality and related conditions to facilitate academic researchers' access to operational networks and real-time data.
- Federal mission and regulatory agencies with primary responsibility for the services industries (e.g., Securities and Exchange Commission, Internal Revenue Service, Federal Communications Commission, and U.S. Department of Health and Human Services) should consider funding academic research in ways that encourage greater participation by the services industries. Engineering research centers funded by the National

Science Foundation and university transportation centers funded by the U.S. Department of Transportation could serve as models.

General Recommendation 4. Individual researchers and organizations, such as the Association of University Technology Managers, that gather data on university research and technology-transfer activities should continue to monitor and assess the effectiveness of incentives for transferring academic research results (particularly intellectual property policies and practices) and the impact of entrepreneurial activity by academic researchers on the traditional university missions of education, research, and service. The following issues should be addressed:

- The costs to institutions of patenting research results, including the costs of maintaining and defending patents, should be assessed and compared to the benefits, in terms of income from licenses and royalties.
- Steps being taken to disseminate patent information to improve the chances of commercialization should be reviewed and best practices identified.
- Best practices in the long-term management of patent inventories should be shared among research institutions.
- The effectiveness of technology transfer via patented inventions should be assessed and compared to transfer via more traditional mechanisms, such as publications. The benefits to faculty and universities should also be compared.
- The impact of university-industry research collaboration and technology transfer activities on undergraduate, graduate, and continuing education, the composition of academic research, the stability of academic research funding, the private and social returns from academic research, the many traditional service roles of the university, and other related issues should be assessed.

General Recommendation 5. Government regulatory agencies, including the Food and Drug Administration, the Environmental Protection Agency, the Federal Communications Commission, and the Securities and Exchange Commission, should be encouraged to sustain and strengthen their productive interaction with academic researchers and to continue to explore new mechanisms for bringing scientific and engineering advances, including scientifically based concepts and tools, to bear more rapidly and effectively on regulatory processes.

General Recommendation 6. Government, industry, and universities should work together to meet the challenges and opportunities created by information technologies. The following steps would be beneficial:

- Boost federal funding for fundamental research in information technologies, as part of an effort to redress the imbalance in federal funding for various disciplines in academic research.
- Increase public and private sector investment in software research, with an emphasis on (1) engineering methods for assessing and improving quality and (2) software that is more flexible and responsive to changing business conditions.
- Support more interdisciplinary research on existing and potential information technologies that combines engineering methods and the social and behavioral sciences.

REFERENCE

NRC (National Research Council). 2003. Innovation and Information Technology. Washington, D.C.: The National Academies Press.

1

Introduction

The importance of research universities to the strength of the U.S. innovation system and the national economy has long been recognized. In recent decades, public policies have sought to bolster the contribution of the nation's academic research enterprise to U.S. industry and economic growth. Nevertheless, as of the late 1990s relatively few studies had been done on the contributions and impact of academic research on specific industrial sectors of the economy. To address this gap, the Alfred P. Sloan Foundation sponsored a number of industry-specific studies in the late 1990s. In 1998, as part of that series, the National Academy of Engineering (NAE) was asked to undertake this study to assess the contributions of academic research to five high-technology manufacturing and service industries reflecting the diversity and shifting balances in the national economy. The five industries are: network systems and communications; medical devices and equipment; aerospace; transportation, distribution, and logistics services; and financial services.

This report, based on the deliberations of the study committee and the fact-finding activities of industry-specific panels of experts, describes specific contributions of academic research to these industries and modes of interaction between industries and universities. It also provides qualitative assessments of the impact of academic research on the performance of each industry and, where applicable, identifies trends and emerging opportunities for increasing contributions from academic research.

SETTING THE CONTEXT

The critical role of university-based research and associated activities in the U.S. system of technological innovation is well documented (e.g., Brooks and Randazzese, 1998; Cohen et al., 1998; Florida and Cohen, 1999; *Journal of Technology Transfer,* 2001; and Mowery et al., 1999). With the increase in government support after World War II, university research flourished. In the 1970s, research and development (R&D) performed by universities comprised around 10 percent of all R&D in the United States. After a slight dip in the early 1980s, universities' share rose to about 13 percent in 2002 (NSB, 2000; NSF, 2003). In 2002, nearly three-fourths of total university R&D expenditures were classified as basic research and about 22 percent as applied research (NSF, 2003). Universities account for roughly half of the basic research performed in the United States (NSF, 2003).[1]

Figure 1-1 shows the sources of funding for university R&D from 1953 to 2002. Since 1953, the largest share has been provided by the federal government, although the percentage of federal support fell from a high of more than 70 percent in the mid-1960s to 60 percent in 2002. By contrast, funding from industry fluctuated from 8 percent in the 1950s to a low of 2 percent in 1966 to 7 percent from 1988 to 2001 (NSF, 2003). From 1968 to 2002, however, industry was the fastest growing source of funding for academic R&D. In constant 1996 dollars, industrial support of academic R&D grew nearly 900 percent

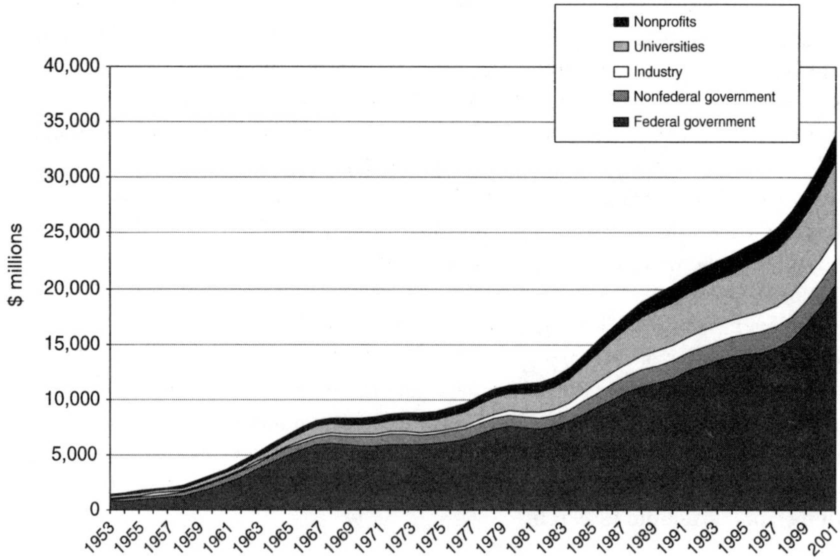

FIGURE 1-1 University R&D funding by source (constant 1996 dollars).
Source: NSF, 2003.

during this period (although it started from a relatively small base) (NSF, 2003). Several surveys from the 1990s indicate that industry accounts for a significantly larger share of funding for academic R&D in specific fields, especially engineering. A 1993 survey of engineering faculty members active in research showed that 17 percent of their research support came from industry (Morgan et al., 1994). A survey of more than 1,000 university research centers in 1990, many of which were founded during the 1980s, revealed that 31 percent of their support came from industry (Cohen et al., 1994).[2]

Figure 1-1 also shows that, since 1968, the second fastest growing source of funding has been internal academic resources, which increased by nearly 700 percent and accounted for 20 percent of R&D funding in universities and colleges by 2002. Internal funds from universities include general-purpose state and local appropriations, general-purpose grants from outside sources, royalty income, endowment income, and unrestricted gifts.

Figure 1-2 shows changes in federal support for academic research by agency. The most striking change since 1970 is the large increase in funding from the National Institutes of Health (NIH); the share of funding from other agencies remained level or declined slightly. This shift has resulted in a corresponding shift in the technical fields supported by federal research funds (Rapoport, 1999). Figure 1-3 shows that, from 1973 to 2001, the percentage of total federal funding for academic research dedicated to the life sciences rose

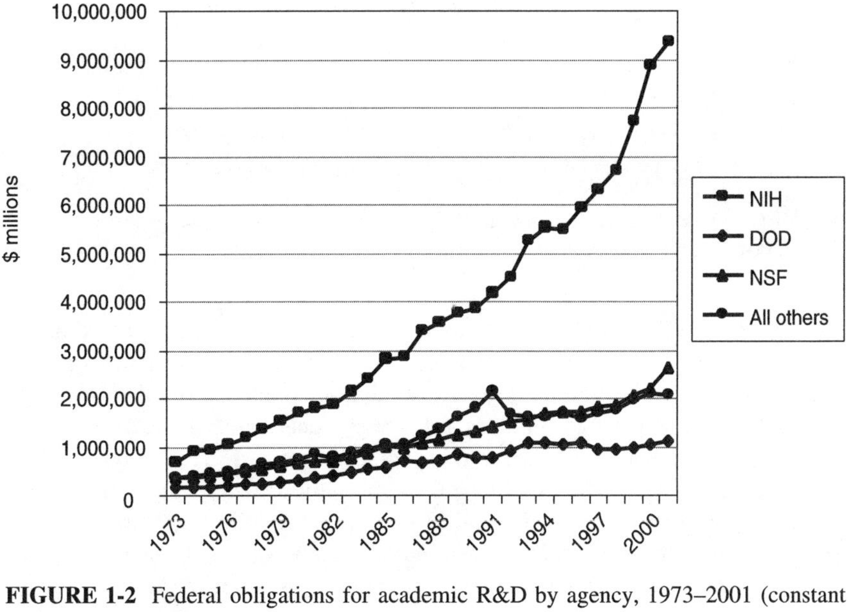

FIGURE 1-2 Federal obligations for academic R&D by agency, 1973–2001 (constant 1992 dollars). Source: NSF, 2001a.

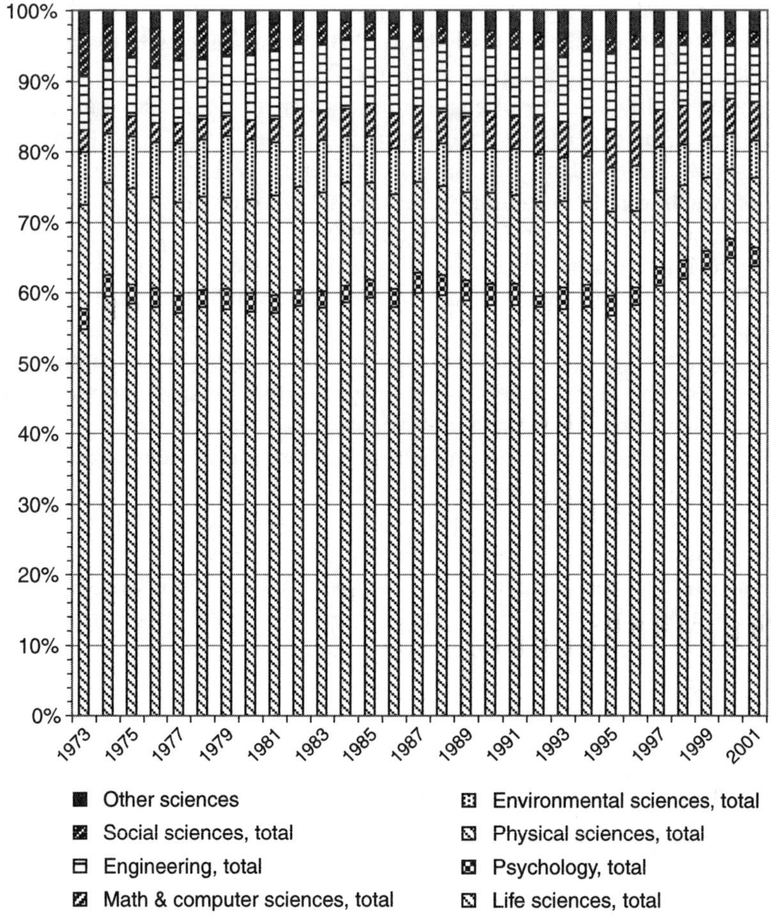

FIGURE 1-3 Percentage of federal obligations for total academic research by field, 1973–2001. Source: NSF, 2001b.

from 54.8 to 63.7 percent, and the share allocated to mathematics and computer science research increased slightly from 3.1 to 5.5 percent. During this period, engineering's share of the total federal funding of academic research remained flat at roughly 8 percent; funding for the physical sciences decreased from 14.7 to 9.9 percent; funding for the social sciences decreased from 5.8 to 1.8 percent (NSF, 2001b). During this same period, federal funding for academic research increased from \$1.6 billion to \$15.2 billion in current dollars (NSF, 2001b). The amount and distribution of support for academic research in specific subfields (e.g., aeronautical engineering vs. biomedical engineering) varies widely over time.[3]

Another indicator of the role of universities in the U.S. research enterprise is patenting. Throughout the 1970s and early 1980s, government policy increasingly favored stronger protection for intellectual property resulting from publicly funded research. Several universities had already increased patenting activity in the 1970s, largely as a result of the emergence of biotechnology, but the propensity to patent increased markedly after 1980.[4] Before 1980, all American universities were issued fewer than 250 patents annually; by the mid-1990s the number had exceeded 1,500, the majority in the health care and life sciences (Nelson, 2000). Overall, the number of university-held patents increased more than 10-fold between 1982 and 1998 (NSB, 2000). In fiscal year 2000 (FY00), U.S. universities applied for more than 6,300 new patents and were issued more than 3,700 (AUTM, 2001). Other indicators of the growing industrial relevance of academic research are the approximate doubling of the percentage of papers coauthored by university and industry personnel from 1981 to 1995 (NSB, 1998) and an increase in the percentage, from 49 percent in 1988 to 55 percent in 1996, of front pages of industrial patents that cite university papers (Jankowski, 1999).

Research by the Association of University Technology Managers (AUTM) shows that university involvement in several kinds of commercially relevant activities accelerated in the 1990s (AUTM, 2000). In FY00, at least 368 new companies were formed based on academic discoveries. More than 100 licenses to university patents generated over $1 million apiece for their university owners, with the University of California system holding 10 of these and Columbia University 16. Total license income to universities in FY00 was $1.076 billion, much of it attributable to a few big winners (AUTM, 2001). Examples include the $160 million earned by Michigan State University over the life of two cancer-related patents, the $143 million lifetime earnings of Stanford University for the recombinant DNA gene-splicing patent, the $27 million paid to Iowa State University for the fax algorithm, and the $37 million earned by the University of Florida for Gatorade (Rogers et al., 2000).

The combination of growth in nonfederal funding sources, shifts in the technical fields receiving the largest share of government support, the increase in academic ownership of intellectual property resulting from academic research, and the increase in financial rewards from the licensing and commercialization of research results have changed the role of academic research in the U.S. innovation system. The university's role of participating in large government research programs driven by Cold War priorities has given way to an environment in which university researchers and research managers are much more aware of the potential for the commercialization of their research. Greater ownership, and therefore greater control, of intellectual property has put research universities more directly into the value chain of new technology development, especially in the life sciences. Technology transfer offices have been created at more than 200 research universities to manage patents, licensing and royalty contracts, and other business and legal issues related to universities' intellectual property

(Sampat and Nelson, 1999). Although the flow of license and royalty income remains confined to a small number of universities—just 10 universities plus the University of California system accounted for more than two-thirds of license income in FY00—university managers now recognize the potential of generating income from their research and are paying much more attention to increasing returns on their intellectual property by managing relationships with corporations, supporting start-up companies by faculty, and negotiating beneficial licensing agreements and royalty terms (AUTM, 2001).

ASSESSING THE IMPACTS

If the economic benefits of academic research could be unambiguously quantified, the choice of effective research investment policies might be straightforward. Unfortunately, it is extremely difficult to apportion the economic output of an industry to the results of academic research. Although a large number of specific results of university research—certain drugs, software applications, algorithms—have been applied directly to industrial practice and products, the overall benefits of academic research are surely much larger. The cumulative scientific and technological knowledge in a field that underlies varied products and services may be even more important than specific inventions or innovations. A programmer who writes a Web-based order-entry system for a new business, for example, makes use of countless contributions from academic research that led to low-cost integrated circuit chips, software tools, the Internet, and a myriad of related technologies.

Several economists have attempted to calculate the private returns (returns to investing firms) and the social returns (returns to both firms and consumers) from university research.[5] Others have used regression analysis to evaluate the impact of academic research on industrial patenting, manufacturing productivity, and other measures of industrial performance (Adams, 1990; Jaffe, 1989). Possible approaches to a quantitative assessment of the impact of academic research on industrial performance are summarized in Box 1-1.

No matter which approach is adopted, measuring the impact of academic research is an inexact science. Isolating, tracking, and measuring the contributions of a given body of academic research to the performance of particular firms, industries, and regional economies is complex and difficult (Feller, 1996). Therefore, this committee's assessments of individual industries are necessarily qualitative, based on informed judgments by knowledgeable experts in industry and universities, as well as case studies, informal surveys, workshop deliberations, and examples.

The NAE deliberately selected diverse industries for this study. Each industry is influenced by many factors, such as size, age, trade dependence, and research intensity, all of which influence both the amount of relevant research performed and the ability of an industry to absorb the results. On the one hand,

Box 1-1
Quantitative Measurements of the Impact of Academic Research

Patents and patent citations could be used as proxies for industrial performance to trace the contributions of academic research to patents. Francis Narin and colleagues at CHI Research, Inc., identified a sizable increase in the citation linkages between U.S. patents and scientific research papers from the late 1980s to the mid-1990s. "References from U.S. patents to U.S. authored research papers have tripled over a six year period, from 17,000 during 1987–1988 to 50,000 during 1993–1994, a period in which the U.S. patent system grew only by 30 percent." The authors concluded that "public science plays an essential role in supporting U.S. industry, across all the science-linked areas of industry, amongst companies large and small, and is a fundamental pillar of the advance of U.S. technology . . . [furthermore] the data show that the science that is contributing to high technology is mainstream; it is quite basic, quite recent, published in highly influential journals, authored at major universities and laboratories, and supported by NSF, NIH, the Departments of Defense and Energy, and by other public and charitable institutions" (Narin et al., 1998). Although this study did not focus specifically on universities, the authors also found that about half of the U.S. papers cited in U.S. industry patents were authored at academic institutions.

Porter and Stern (1998) used international patents per million persons as an output measure of innovation, and correlated it with other inputs, such as investments in education and R&D. Perhaps academic patents could be used in this fashion.

Surveys

Survey research could be used to obtain information about a variety of indicators of university research and industrial performance. Based partly on a 1994 survey of industrial R&D managers, W. Cohen and associates at Carnegie Mellon University concluded that "university research provides critical short-term payoffs in some industries (such as pharmaceuticals) and is broadly important in numerous industries" (Cohen et al., 1998). The survey provided information about the percentage of industry R&D projects using university research and the extent to which industries used various university research results, including patents, for moving information from universities to industrial R&D facilities.

Morgan, Strickland, and colleagues conducted mail surveys and telephone interviews with engineering and science faculty about research they conducted that they believe contributed to industry (Morgan et al., 1997). Interviews were also conducted with industry counterparts or collaborators. The authors concluded that university engineering and science research is often relevant to industry and that real, tangible contributions are being made. However, when interviewees were asked to place a dollar value on their contributions, relatively few responded (Morgan et al., 1997).

Quantifying Dollar Values

Overall impacts of individual universities. A study released in 1997 attempting to measure the impact of one research university, MIT, on the economy and employment concluded, "If the companies founded by MIT graduates and faculty formed an independent nation, the revenues produced by the companies would make that nation the 24th largest economy in the world. The 4,000 MIT-related companies employ 1.1 million people and have annual world sales of $232 billion." MIT-

(continued)

related firms are prominent in electronics, software, and biotechnology and tend to be concentrated in Silicon Valley and Boston. More than half of the companies founded by MIT graduates were founded by graduates in electrical engineering (BankBoston, 1997).

In a study conducted in the early 1990s, the Stanford University licensing office compiled information about technology-based companies founded by members of the Stanford community (graduates, faculty, etc.). An aggregate estimate of roughly $31 billion in revenue was attributed to firms in the San Francisco Bay area (Leone et al., undated).[1]

Dollar impacts of university patents and licenses. Using economic impact models, the Association of University Technology Managers (AUTM) estimates that in FY98, $33 billion of U.S. economic activity was attributable to the results of academic licensing from U.S. universities that participate in the AUTM survey, supporting 280,000 jobs (AUTM, 1999). In another study, Pressman (2000) reports that 150 licensees with agreements to use MIT-owned patents sold $3 billion in licensed products. More than 500 of some 850 MIT patent license agreements since 1980 are active.

Estimating Economic and Social Returns

Mansfield estimated the annual social returns from academic research, which include returns to both firms and consumers, from 1975 through 1978 to be 28 percent, although he warned that the estimate was very rough. Mansfield also estimated that sales of new products in 1985, based on academic research that was first commercialized between 1982 and 1985 in seven industries, including drugs, instruments, and information processing, totaled about $41 billion, or 5 percent of total sales for those industries (Mansfield, 1991). Additional benefits were estimated for process improvements. Mansfield's work was based in part on surveys of firms that indicated that without university research, about 10 percent of industrial innovations would not have occurred or would have occurred only after sizable delays.

Trajtenberg (1999) estimated the social return of computerized tomography scanners to be more than 200 percent. A question that then arises is what portion of the returns from this innovation can be attributed to academic research. Given that fundamental scientific discoveries provided the basis for producing CT scanners, direct dependencies are difficult to discern; thus, the contribution of academic research is also extremely difficult, if not impossible, to quantify.

Cohen et al. (1998) summarized studies by Jaffe (1989) and Adams (1990) in which regression analyses were used along with production functions to evaluate the impact of academic research on industry. Jaffe's (1989) study was focused on the effects of academic R&D on industrial patenting, whereas Adams (1990) examined effects on manufacturing productivity.

Using Multiple Approaches and Proxies

Jaffe (1998) suggested a combined approach to quantifying the impacts of academic research on industrial performance. "I do not believe that it is possible to perform a reliable and comprehensive measurement of the outcomes of science and technology programs. This does not mean, however, that measurement is pointless. By looking at multiple indicators, captured and evaluated at multiple levels of aggregation (individual projects, individual programs, agencies, and the economy as a whole) we can draw reasonably reliable conclusions. Further, the use of multiple indicators reduces the need to be overly concerned about the limitations of any one."

[1] MIT and Stanford may be counting some of the same people (e.g., someone with a B.S. from MIT and a Ph.D. from Stanford or vice versa).

the diversity of the five industries precludes the emergence of universal, economy-wide truths. On the other hand, the committee was able to identify best practices that could be generally implemented, as well as critical issues and shortcomings common to these five industries, and perhaps others.

For example, the studies of the five industries clearly indicate the importance of research in both the natural sciences and engineering, as well as the social and behavioral sciences, in the development and broad implementation of innovations. Academic studies of consumer behavior, for instance, have been important to the emergence and success of electronic commerce. Policy research on regulations and standards has contributed to changes in banking (the end of Glass-Steagall, which prohibited banks from accepting deposits and underwriting securities and established the Federal Deposit Insurance Corporation), communications (the break-up of AT&T), logistics (the deregulation of trucking and airlines), and a wide range of similar legal and regulatory changes. The contributions of academic research to a social, legal, political, and economic environment that encourages innovation are often overshadowed by the more familiar contributions to advances in science and technology.

SECTOR-SPECIFIC STUDIES

This study is based on the premise that the research and technology needs and strategies of industries are sector specific as well as company specific, as is the character of academic-industry interactions. Because the frontiers of commercial research, development, and innovation vary widely among companies in the same industry and among industries, research and resulting innovations can have a wide range of effects on commercial success. Some of the differences are a function of the characteristics of science and technology, but most of them are attributable to differences in research intensity, necessary returns on investment, market size, competition, product mix, and other distinctive characteristics of an industry. The five industries examined in this study encompass a great many variations.

Box 1-2 summarizes the five industries, all of which are important in terms of U.S. sales and employment, technological intensity, expected growth rates, and other metrics. One of the difficulties encountered in defining these industries is that subsectors in an industry may differ widely. In network systems and communications, the panel focused on the innovative computer networking segment of the industry rather than on older communications services, such as telephony. In the medical devices and equipment industry, the panel assessed only the more sophisticated subsectors of the industry. The aerospace panel focused on five subsectors of the aerospace industry, three of which are large, mature businesses whose interactions with universities differ substantially from those of the two smaller subsectors. In transportation, distribution, and logistics services, the panel focused primarily on the contributions of one technologically

Box 1-2
Industry Definitions

Network Systems and Communications. Focused was on the innovative computer networking segment of the industry, including computing and communications equipment, software, and services (basically all information technology except for integrated circuits). The industry generated roughly $715 billion in sales and more than 2.2 million jobs in 2000.

Medical Devices and Equipment. Includes five North American Industrial Classification System (NAICS) codes: (1) surgical and medical instruments, (2) surgical appliances and supplies, (3) dental equipment and supplies, (4) irradiation apparatus, (5) and navigational, measuring, electromedical, and control instruments. Medical information systems are also included. As of 1999, the global market for medical devices was $138 billion, and the U.S. market accounted for 37 percent of global demand. The U.S. industry supplied 40 percent of the global market with shipments of $55 billion and employed nearly 300,000 workers.

Aerospace. Includes five sectors, ranging from the mature (1) large-scale airframes, (2) jet engine, and (3) launch-vehicle businesses to the newer businesses of (4) unmanned aerial vehicles and (5) space-based information systems. The industry as a whole accounted for more than $146 billion in sales and 800,000 jobs in 2000.

Transportation, Distribution, and Logistics Services. Focuses on integrated logistics services, including inventory carrying costs and transportation and administrative costs associated with moving freight and people. The industry had roughly $1 trillion in sales and employed more than 3.6 million people in 2000.

Financial Services. Includes all services associated with the packaging and description of financial securities and the implementation of financial transactions. The industry accounted for roughly $820 billion of U.S. gross domestic product and provided nearly 6 million jobs in 2000.

Sources: AIA, 2001; Delaney and Wilson, 2001; McGraw-Hill and U.S. Department of Commerce, 2000; U.S. Bureau of the Census, 2001, 2002.

driven approach, namely integrated logistics, to the movement of freight. This subsector has undergone a good deal of innovation with significant contributions from academic research; the industry as a whole, however, has not. The financial services panel took a comprehensive view of the industry, including consumer and commercial services, investment banking, and insurance. The medical devices and equipment industry and the network systems and communications industry have a history of extensive collaboration with academic research. The aerospace industry, the financial services industry, and the transportation, distribution and logistics services industry generally have had more limited, less systematic interactions with academic researchers.

All five industries have demonstrated an impressive capacity for innovation over the past decade. However, only the three manufacturing industries (network systems and communications; medical devices; and aerospace) have long-standing, well developed R&D functions. By contrast, the two predominantly service industries (transportation, distribution, and logistics services; and financial services) have only recently begun to develop an R&D ethos in the traditional sense. The following chapters describe each industry in greater detail and summarize the findings for each industry.

NOTES

[1]There are serious limitations to using the categories basic research, applied research, and development to characterize the R&D enterprise (Stokes, 1997). In spite of the uncertainty in placing specific research projects into these categories, science and technology agencies that produce R&D data routinely use them.

[2]In some cases, the federal government requires that a university obtain matching funds from industry as a prerequisite for supporting a research center.

[3]The five industry sectors examined in this study do not correspond exactly to categories for which industry research and other data are collected. Furthermore, academic research in several fields may contribute to these industries, making it difficult to assign the contributions of specific disciplinary research in universities to a specific industry sector.

[4]Several factors contributed to this: a 1980 Supreme Court decision, *Diamond vs Chakrabarty*, upheld the validity of a broad patent in biotechnology; the Patent and Trademarks Law Amendments Act (P.L. 96-517), often referred to as the Bayh-Dole Act of 1980, changed long-standing federal policy that had not allowed research institutions to own inventions developed under federal research contracts or to license or otherwise pursue commercialization of the invention; and the creation of the Federal Circuit Court of Appeals in 1982 provided a strong champion of patent-holder rights (Mowery et al., 1999).

[5]In an assessment of seven industries, Mansfield (1991) estimated that the annual social returns from academic research from 1975 to 1978 were 28 percent. Mansfield also estimated that 1985 sales of new products based on academic research that were first commercialized between 1982 and 1985 totaled about $41 billion, or 5 percent of total sales for those industries. For a review of approaches to measuring social returns of R&D not specifically focused on academic research, see Jones and Williams, 1998.

REFERENCES

Adams, J. 1990. Fundamental stocks of knowledge and productivity growth. Journal of Political Economy 98(4): 673–702.

AIA (Aerospace Industries Association). 2001, Aerospace Facts and Figures: 2001/2002. Washington, D.C.: Aerospace Industries Association. Available online at: *http://www.aia-aerospace.org/stats/facts_figures/ff_01_02/ff_summary.cfm.* [August 21, 2003]

AUTM (Association of University Technology Managers). 1999. FY 98 Licensing Survey. Available online at: *http://www.autm.net/index_ie.html.* [June 24, 2003]

AUTM. 2000. AUTM Licensing Survey, FY 1999. Northbrook, Ill.: AUTM.

AUTM. 2001. AUTM Licensing Survey, FY 2000. Northbrook, Ill.: AUTM.

BankBoston. 1997. MIT: The Impact of Innovation. Special Report of the BankBoston Economics Department. March. Available online at: *http://web.mit.edu/newsoffice/founders/.* [June 24, 2003]

Brooks, H., and L.P. Randazzese. 1998. University-Industry Relations: The Next Four Years and Beyond. Pp. 361–399 in Investing in Innovation: Creating a Research and Innovation Policy that Works, L.M. Branscomb and J.H. Keller, eds. Cambridge, Mass.: MIT Press.

Cohen, W.M., R. Florida, and W.R. Goe. 1994. University-Industry Research Centers in the United States. Pittsburgh, Pa.: Center for Economic Development, Carnegie Mellon University.

Cohen, W.M., R. Florida, L.P. Randazzese, and J. Walsh. 1998. Industry and the Academy: Uneasy Partners in the Cause of Technological Advance. Pp. 171–199 in Challenges to Research Universities, R. Noll, ed. Washington, D.C.: The Brookings Institution.

Delaney and Wilson. 2001. Managing Logistics in a Perfect Storm. 12th Annual State of Logistics Report. Washington, D.C.: Cass Information Systems.

Feller, I. 1996. Technology Transfer from Universities. Pp. 1–42 in Higher Education: Handbook of Theory and Research, J.C. Smart, ed. New York: Agathon Press.

Florida, R., and W.M. Cohen. 1999. Engine or Infrastructure? The University Role in Economic Development. Pp. 589–610 in Industrializing Knowledge, L.M. Branscomb, F. Kodama, and R. Florida, eds. Cambridge, Mass.: MIT Press.

Jaffe, A. 1989. Real effects of academic research. American Economic Review 79(5): 957–970.

Jankowski, J.E. 1999. Trends in academic research spending, alliances, and commercialization. Journal of Technology Transfer 24: 55–68.

Jones, C.I., and J.C. Williams. 1998. Measuring the social return to R&D. Quarterly Journal of Economics 113(4): 1119–1127.

Journal of Technology Transfer. 2001. Symposium on Technology Organizational Issues in University-Industry Technology Transfer. Special Issue. Journal of Technology Transfer 26(1).

Leone, A., J. Vamos, R. Keeley, and W.F. Miller. Undated. A Survey of Technology Based Companies Founded by Members of the Stanford University Community. An unpublished study commissioned by the Stanford University Office of Licensing, Stanford, California.

Mansfield, E. 1991. Academic research and industrial innovation. Research Policy 20: 1–12.

McGraw-Hill and U.S. Department of Commerce. 2000. Chapter 45, Medical and Dental Instruments and Supplies, in U.S. Industry and Trade Outlook 1999. New York: McGraw-Hill.

Morgan, R.P., D.E. Strickland, N. Kannankutty, and J. Grillon. 1994. Research on academic engineering research. Part II. How engineering faculty view academic research. ASEE PRISM 4(3): 30–35.

Morgan, R.P., D.E. Strickland, M.E. Sava, and N. Kannankutty. 1997. Academic Engineering Research at a Time of Change: The Tilt Towards Industry. Pp. 226–234 in Proceedings of IEEE International Symposium on Technology and Society. New York: IEEE.

Mowery, D.C., R.R. Nelson, B.N. Sampat, and A.A. Ziedonis. 1999. The Effects of the Bayh-Dole Act on U.S. Research and Technology. Pp. 269–306 in Industrializing Knowledge, L.M. Branscomb, F. Kodama, and R. Florida, eds. Cambridge, Mass.: MIT Press.

Narin, F., K. Hamilton, and D. Olivastro. 1998. The Increasing Linkage between U.S. Technology and Public Science. P. 101 in AAAS 1998 Science and Technology Yearbook, A. Teich, S.D. Nelson, and C. McEnaney, editors. Washington, D.C.: AAAS.

Nelson, L. 2000. Many Forms of Technology Transfer from Universities. Viewgraph presented at the AAAS Science and Technology Policy Colloquium, April 2000, Washington, D.C.

NSB (National Science Board). 1998. Science and Engineering Indicators 1998. Washington, D.C: U.S. Government Printing Office.

NSB. 2000. Science and Engineering Indicators 2000. Washington, D.C.: U.S. Government Printing Office.

NSF (National Science Foundation). 2001a. Federal Funds for Research and Development: Detailed Historical Tables: Fiscal Years 1951–2001. NSF 01-334. Arlington, Va.: NSF.

NSF. 2001b. Federal Funds for Research and Development: Federal Obligations for Research to Universities and Colleges by Agency and Detailed Field of Science and Engineering: Fiscal years 1973–2001. NSF 01-331. Arlington, Va.: National Science Foundation.

NSF. 2003. National Patterns of Research and Development Resources: 2002 Data Update. Arlington, Va.: National Science Foundation. NSF 03-313. Available online at: *http://www.nsf.gov/sbe/srs/nsf03313/start.htm. [June 24, 2003]*

Porter, M., and S. Stern. 1998. Evaluating U.S. Innovative Capacity: The Innovation Index. Presentation at MIT Innovation Summit, March 13, 1998, Cambridge, Massachusetts.

Pressman, L. 2000. Measuring Product Development Outcomes of Patent Licensing at MIT. Paper presented at AAAS Annual Meeting, Washington, D.C., February 17–22, 2000.

Rapoport, A.I. 1999. How Has the Field Mix of Federal Research Funding Changed over the Past Three Decades? Issue Brief NSF 99-328. Arlington, Va.: National Science Foundation, Division of Science Resources Studies.

Rogers, E.M., J. Yin, and J. Hoffman. 2000. Assessing the effectiveness of technology transfer offices at U.S. research universities. Journal of the Association of University Technology Managers 12: 47–80.

Sampat, B., and R. Nelson. 1999. The Emergence of Standardization of University Technology Transfer Offices: A Case Study of Institutional Change. Prepared for the New Institutional Economics, September 16–18, 1999. Washington, D.C.: World Bank.

Stokes, D.E. 1997. Pasteur's Quadrant: Basic Science and Technological Innovation. Washington, D.C.: The Brookings Institution.

Trajtenberg, M. 1999. Economic Analysis of Product Innovation: The Case of CT Scanners. Cambridge, Mass.: Harvard University Press.

U.S. Bureau of the Census. 2001. Statistical Abstract of the United States. Washington, D.C.: U.S. Government Printing Office.

U.S. Bureau of the Census. 2002. Statistical Abstract of the United States–2002. Washington, D.C.: U.S. Government Printing Office. Available online at: *http://www.census.gov/prod/www/statistical-abstract-02.html.* [June 24, 2003]

2

Report of the Panel on the Network Systems and Communications Industry

The Panel on Network Systems and Communications, one of five panels formed by the Committee on the Impact of Academic Research on Industrial Performance, was asked to examine the impact of academic research on the performance of the network systems and communications industry and recommend ways based on trends in the industry and the research community to increase this impact. The panel of six included three members of NAE (all from industry), one other member from industry, and two from academia. Three of the panel members were also members of the parent committee. The panel reviewed the literature, developed several case studies, and sent a questionnaire to experts in academia, the computer industry, the network systems and communications industry, and government. The questionnaire was followed by a workshop attended by approximately 30 senior individuals in the network systems and communications sector (see Addendum).

The network systems and communications business sector flourished throughout the 1990s, when the growth of the Internet, the technologies that implement it, and the businesses and services that use it were unprecedented. Telecommunications services—especially wireless digital telephones and paging services—also grew rapidly. Much of this success was attributable to exponential improvements in the performance-to-cost ratio of microelectronics over the past three decades. Technical innovations emerging from within the industry and from academic research have been essential. Some innovations were the culmination of decades of research; some were short-term developments that entered the market via start-up companies; and some were incremental improvements to existing products.

In the last 30 years, digital technologies have transformed the telephone network from an analog system to a computer-controlled system with digital switching and transmission. The process of digitalization has changed the industry from two distinct businesses—computers and communications—to one business in which computers and communications are intermingled in products and services. This convergence was accelerated by advances in microelectronics and increases in the bandwidth available for communications (Messerschmitt, 1996). The result is increasingly pervasive data networking, based largely on the packet-switching technologies that emerged from academic and industrial research to spawn the Internet.

The network systems and communications industry has a large and expanding services component. For the telecommunications industry, which has always been a service provider, the challenge is to invent and offer customers new, valuable services that generate new sources of revenue. For the computer portion of the industry, high-performance communications are making a wide range of new services feasible. Examples include remote sensors and control systems; integrated supply chain management systems; application service providers; full-time, real-time stock quotes; and instant messaging.

DEFINITION OF THE INDUSTRY

The network systems and communications industry must be defined very broadly. It clearly includes the manufacturing of telecommunications equipment and the services that use such equipment, such as telephony, wireless telephony, broadcast television, cable and satellite television, radio, and Internet service. Both the equipment and services sectors increasingly require computing equipment and software, and, in fact, the computer and communications industries are no longer separate industries. For example, cellular telephony depends on a broad range of technologies: the cell phone contains a liquid crystal display, an embedded computer with a lot of software, and advanced chips that integrate most of the components of a high-frequency radio; the transmission formats depend on advances in digital speech compression, signal modulation, and coding; the base stations depend heavily on digital integrated circuits and computers for switching and control and fiber-optic links between them; tracking a moving telephone requires that computers at adjacent base stations exchange protocol messages for the handoff; and the billing, provisioning, and maintenance of the service require large-scale computing and software systems of the service provider. Separating this integrated system into "communications" and "computing" components is simply not possible. In short, computing and communications equipment and services have converged, creating new business and technical opportunities.

The explosive growth of the Internet is the most visible manifestation of this trend toward convergence. The technologies underlying the Internet—just like those that underlie the cellular telephone—include computing and communications. Special computers serve as routers, and network services knit together the transmission links and implement the collection of Internet protocols that carry Internet traffic. The explosive growth of the Internet, however, is attributable not to these basic provisions—which existed before 1993—but to new services that created consumer demand: electronic mail, the World Wide Web of information and its associated browser software; chat groups; real-time delivery of audio and video media; online merchandising; banking and financial transactions; supply-chain integration of suppliers and customers; and numerous other applications. Some applications merely extend existing internal information technology systems to provide Internet access. But others, such as eBay's success with online auctions, are entirely new business concepts. As the Internet becomes more pervasive, old ways of computing, in which data was created, stored, and manipulated at a single site, are giving way to networked systems in which data can be accessed remotely and shared extensively.

The computers embedded in everyday objects—telephones, cars, televisions, furnaces, hi-fi equipment—are becoming increasingly capable and increasingly networked. Some cars already can connect with a diagnostic and help center by cellular telephone or satellite communications. Home networks in which multiple personal computers in a household are linked over existing telephone wires and short-range wireless devices will soon make networking of appliances routine. A world in which all devices have an Internet address is not out of reach. Thus computers increasingly require communications to fulfill their functions, and communications increasingly require computers to fulfill theirs.

The technologies of computing and communications are becoming indistinguishable. All of them depend on software to express functions at all levels in the network. A few years ago, a modem was a complex, integrated circuit. Today, with more complex algorithms and faster computers, modems are written in software embedded within digital signal processors. Many algorithms can be used equally well in computing and communications settings. For example, schemes to digitize and compress video signals are useful both for manipulating and storing video information on a computer disk and for transmitting it over digital communications channels. Similarly, encryption technology can be used to protect sensitive information in a computer system or in transit over a network.

These three trends—convergence, embedding, and network applications—characterize the network systems and communications sector. The panel's assessment of the contributions of academic research to this industry is based on this broad definition.

TABLE 2-1 Sales and Employment in the Information Technology Industry, 2000

	NAICS[a] Code	Sales Revenues ($ billions)	Number of Jobs (1,000)
IT Manufacturing			
Computer and peripheral equipment	3341	$110.0	190
Communications equipment	3342	119.3	291
Software	5112	88.6	331
Semiconductors and other electronic components	3344	168.5	621
IT Services			
Data processing services	5142	42.9	296
Telecommunications services	5133	354.2	1,165

[a]North American Industrial Classification System. Source: U.S. Bureau of the Census, 2002.

Size

Because our definition has vague boundaries and because the industrial classifications used to gather statistics have not been adapted to the rapid changes in the industry, it is difficult to determine the size of the network systems and communications sector. Table 2-1 summarizes sales and employment in the information technology industry based on Bureau of the Census data (U.S. Bureau of the Census, 2002). Taken together, sales of computer and communications equipment and services (all information technology minus semiconductors) were about $715 billion in 2000, and the industry employed more than 2.2 million people (U.S. Bureau of the Census, 2002). Expenditures for information-processing equipment increased almost 10 percent per year on average from 1970 to 1994; the corresponding figure for computers and peripherals was 27.5 percent (NRC, 1999). A 1999 survey found that telecommunications manufacturing was growing by 16.3 percent annually, computer software by 16.6 percent, and computer hardware by 9.5 percent (CTIS, 1999), however these rates have dropped significantly since early 2000.

Structure

The role of research and innovation in the network systems and communications sector can best be understood in the context of the structure of the industry, which influences the mechanisms of innovation and thus how new technologies and products are introduced. The very general description that follows is intended only to reveal similarities and differences with the other industries studied in this report.

Manufacturing

The structure of the computer industry is horizontal; the communications industry was vertically integrated but has been rapidly changing to a horizontal structure as well. In a horizontal structure, numerous suppliers manufacture parts and components that many integrators assemble into subassemblies that are then assembled into final products by numerous competing original equipment manufacturers. The multiplicity of companies at each manufacturing step ensures intense competition throughout the production process, not only in terms of price but also on a wide rage of performance characteristics. For example, manufacturers of personal computers buy disk drives from any one of about a dozen suppliers. A company that needs a customized integrated circuit may design the circuit but use one of several competing semiconductor foundries to manufacture it. Specialized circuit board assembly firms can assemble and test complete circuit boards, giving an electronic design firm the ability to design and sell a unique computer interface board with custom chips without having to invest in either chip or board manufacturing facilities.

The divestiture of AT&T and the subsequent deregulation of communications services forced the communications industry to change from a vertical to a horizontal structure. Today there are many vendors of telecommunications equipment and components. Custom integrated circuits can implement very complex communications functions; coupled with custom-built and proprietary software designs, equipment vendors compete intensely in terms of technology, reliability, and cost of ownership.

Another important feature of the network systems and communications sector is its reliance on components with well defined interfaces. Integrated circuits are a good example: the physical, electrical, and logical behaviors of chip interfaces are specified by the manufacturer and used by the customer to determine how to incorporate a chip into a subsystem with other components. Subsystems then become components of still larger systems. Software, another component, plugs into the operating system that supports it by linking the software interfaces (sometimes called application programming interfaces, or APIs). A piece of software that is compatible with a certain operating system adheres to the interfaces provided by the operating system. Computer systems are built from complex hierarchical assemblies of subsystems and components, sometimes hundreds or even thousands of them. Some of the components are custom built, and some are standard. Thus, interfaces give rise to components, which in turn give rise to businesses structured around buying and selling components.

Key component interfaces become industry standards, which are usually adopted by industry groups to hasten the spread of a new technology, increase sales volume, and, therefore, decrease cost. Standards maintained by a group with broad representation from competitive suppliers are said to be open standards. For example, the Personal Computer Memory Card International Association is an industry group that establishes standards for interchangeable interface cards for laptop computers. The standards group includes several producers of cards and several producers of laptops to ensure that the standard cannot be manipulated to benefit one competitor over another. By contrast, standards promulgated by a single vendor are said to be proprietary. For example, the programming interface for Microsoft's Windows operating system is proprietary; Microsoft specifies it and can change it at will.

Standards play a special role in communications. Broadly speaking, they are necessary to ensure that components and subsystems connected via a communications channel can operate together (e.g., they obey the same conventions for encoding voice signals, multiplexing many simultaneous phone calls on a single channel, performing operation and maintenance functions). Standards of this kind are necessary for guaranteed, sustained interoperability, and changes must be carefully designed to avoid even slight interruptions of network service. New versions of standards must be designed so they can be introduced incrementally, connect new equipment to old, test new protocols, and so on. The same considerations apply to Internet protocols.

Services

Communication services (e.g., voice and data transmission, switching, and distribution) are a major portion of the network systems and communications industry. The number and structure of telecommunications service providers have been in constant flux since the divestiture of AT&T and the deregulation of local telephone services. First, new companies emerged offering wireless telephone services. Then another group of new companies emerged as Internet service providers.

To increase their revenue, carriers have been developing value-added services, such as voice mail, call forwarding, call waiting, 800 service, electronic mail, and virtual private networking, along with conventional transmission and switching services. Internet service providers provide national and regional portals that offer news, chat rooms, advertising, and direct access to the World Wide Web.

Computing services are also a major element of the industry. System integration, the design and deployment of communications and information systems for large clients, has become a major source of revenue for many equipment vendors. In recent years, an important service has been to implement network capabilities across companies' existing computer systems. In some cases, networking has focused on providing Internet access for employees and customers; in others, the focus has been on the development of internal networks linking production and distribution facilities across the company. So far, neither academic nor industrial research has addressed the problems of service delivery in a structured and sustained manner.

INNOVATION SYSTEM

Most innovations are incremental improvements, such as design refinements, improvements in technology and manufacturing processes, a better understanding of customer needs, and integration of previously separate products. For example, important performance metrics for communications equipment include low power and high density (so that many circuits can be accommodated in the confined spaces of wiring closets, boxes mounted on telephone poles, and even central switching offices). Both power and density can be improved by advances in integrated-circuit technologies, which in turn, derive primarily from incremental improvements in fabrication equipment, processing steps, and materials. Research results may be the basis for some of these improvements, and research has achieved major breakthroughs in these areas; this research is performed or funded by materials, equipment, and microchip fabricators, not by the telecommunications equipment manufacturers (see Box 2-1).

BOX 2-1
The Cellular Telephone

The rapid spread of cellular telephones probably epitomizes the popular conception of innovation. The use of cellular telephony started out slowly but then exploded. There were 11 million subscribers in 1992 and 141 million in 2001. The look, feel, and weight of cell phones have also evolved rapidly.

Although the original cellular concept was developed at AT&T Bell Laboratories in 1946, cellular technology was not the outgrowth of fundamental research on radio-frequency propagation and control. Rather, it was the result of demand-driven technological improvements developed by corporate researchers, primarily at Bell Laboratories and Motorola. The long delay from the initial idea to deployment was principally the result of regulatory and business decisions that de-emphasized the development of cellular technologies. Only a few systems, such as improved mobile telephone service, were deployed at all (in 1964). Because of the delay, developers benefited from the microelectronics revolution and were able to use inexpensive microprocessors and integrated circuits to make equipment cheaper, lighter, and less power hungry.

Until recently, cellular telephony in the United States was dominated by the advanced mobile phone service (AMPS) analog standard developed in the early 1970s. The switch to digital transmission occurred more rapidly in Europe, with the GSM (global system for mobile communication) standard, which uses time-division multiple access. A competing digital scheme that relies on spread-spectrum technology (CDMA—code division multiple access) was developed as a proprietary standard by Qualcomm. In March 1999, the firms competing over CDMA intellectual property and products agreed to support a single worldwide standard to promote widespread adoption. Personal communications systems services in the United States use these techniques in a new, larger frequency band. Many cellular telephones contain electronics that will work with more than one of these standards and thus can operate in many areas around the world.

Academic research has played only a small role in the development of these innovations. Until the telephone industry was deregulated, using public funds to compete with Bell Laboratories research was considered pointless. Moreover, Bell Laboratories did not fund academic research in this area; instead they recruited graduates of broad science and engineering programs and trained them in specialized research areas on the job. After the breakup of the Bell system, new firms entered the telecommunications arena, demand for engineers trained in cellular technologies grew, and academia began to respond. Today, considerable academic research is being done in these technologies.

Source: CTIA, 2003; Qualcomm Corporation, 1999; Roessner et al., 1998.

For many businesses, vendors of materials, products, and services throughout the supply chain are major sources of innovation. Buying an integrated-circuit chip, for example, implicitly buys a share of the dramatic improvements in price and performance of integrated circuits (Moore's Law).[1] Over time, innovations will make the chip faster or cheaper or more capable. A telecommunications carrier that wants to deploy Synchronous Optical Network (SONET), a transmission protocol that defines optical carrier levels and their electrically equivalent synchronous transport signals, can purchase switches, multiplexers, and test equipment from the vendors who developed SONET technology. This pattern is a direct consequence of the "horizontal" structure of the industry. Dell Computer, for example, does not have in-house R&D; in effect, Dell is a broker that negotiates attractive deals to buy components and computer-assembly services for its build-a-computer-to-order business. Dell depends on R&D investments by its vendors, especially Intel and Microsoft, that make the microprocessors and operating system software on which the personal computer business depends. Dell's innovations have been in its business model and supply-chain management, not in its technology.

Innovation can also be purchased by acquiring other companies, especially venture-capital-backed start-up companies that have introduced new products with new technologies. A start-up company is a new business, often with an innovative technology but with considerable risk. Often the innovative technology has its origins in academic research. If the company makes good progress, both in technology and in the market (e.g., beta testing, or success in getting its approach adopted by standards consortia), it becomes an attractive target for a larger company seeking to strengthen its technology or product line. For example, Texas Instruments bought Amati; Fore Systems bought Berkeley Networks and Marconi bought Fore Systems; Cisco bought Granite Systems; and Broadcom bought Epigram. Each of the acquired companies had ventured into a new technical area. Epigram, for example, had devised a way to use home telephone wiring to transmit 10Mb Ethernet traffic and had made progress in standardizing the scheme through the Home Phoneline Networking Alliance. Broadcom, itself an innovative fabless chip company specializing in integrating analog and digital functions of cable and twisted-wire modems, saw buying Epigram as a natural way to enhance its core business.

Although high-tech start-ups seldom do research in the classic sense, many behave much like "applied research" projects in an industrial laboratory. They formulate technically aggressive plans based on established principles to pursue and evaluate; the results of experiments often inform several products. For example, Transwitch attempted to increase the telecommunications protocol processing integrated on a single chip, as well as to partition the chip functions into an "architecture" so that a small number of chip designs could be used to build a wide variety of telecommunications products. Both Amati and Epigram conceived ways of using advanced signal-processing techniques to adapt digital transmission to the characteristics of real-world, twisted-pair copper wires (Amati) or in-house telephone wiring (Epigram). The technology-development activities of these companies are much like those in industrial research, but they are done in a commercial setting and with strong incentives to bring innovations to market rapidly.

Industrial research is concentrated in the laboratories of a few of the largest companies, such as Intel, Microsoft, IBM, Compaq, Lucent, AT&T, Hewlett-Packard, Sun Microsystems, and Xerox. Although many of these firms invest 10 to 15 percent or more of revenues in R&D each year, the vast majority of this is for "development," that is, for the engineering of the next generation of products. Research focused on objectives more than 18 months or one or two product cycles out is estimated to be, at most, 5 percent of that 10 to 15 percent, or far less than 1 percent of revenue.[2] A few large companies eschew research, preferring instead to buy innovative companies (e.g., Cisco). Companies in the services sector, however, generally do not engage in or support research. For example, at MCI, which is generally considered a technology leader, the advanced technology group is primarily concerned with testing new equipment and working with vendors to solve interoperability and operation, administration, and management problems.

Industry research is usually driven by market needs but often includes some fundamental or long-range projects as well. For example, IBM's research on the Internet and electronic commerce includes some long-term work on cryptographic systems for security and authentication. Industrial research often links advanced technologies to emerging product needs. For example, as the Java programming language became popular, industry laboratories at Sun, IBM, and elsewhere launched projects to devise advanced techniques for the compilation, synchronization, and code simplification required for its implementation. Previous research results in these areas had not adequately addressed the needs of the Java language, of today's large memories, or of multiprocessor servers. Some of this research is fundamental in the sense that it can be applied to problems other than Java language implementations. In fact, even though research in engineering fields is usually targeted toward meeting specific engineering needs, the results are often useful for many other applications.

One of the companies' aims in operating research laboratories is to expand their capability for bringing in new ideas and new people (Cohen and Levinthal, 1990). The laboratory is expected to recruit people who cannot be recruited by an engineering organization; it is also expected to interact with the intellectual community by attending conferences, publishing papers, collaborating with universities, or entering partnerships with other companies; and it is intended to counteract the risk inherent in the narrow focus of engineering projects on product development.

A Culture of Innovation

Innovation in the network systems and communications industry can take many paths. Even when research plays an essential role, there is no linear path

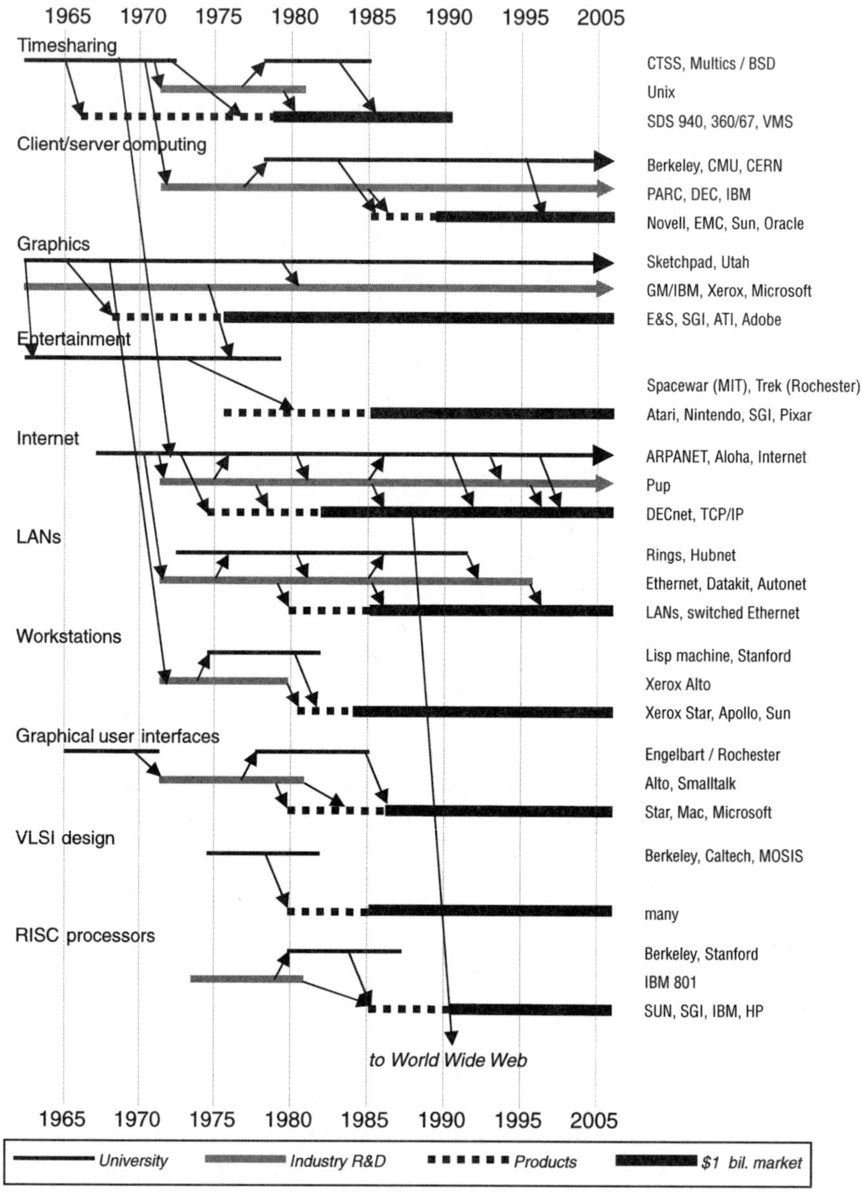

FIGURE 2-1 Examples of academic government-sponsored (and some industry-sponsored) IT research and development in the creation of commercial products and industries. Source: NRC, 2003.

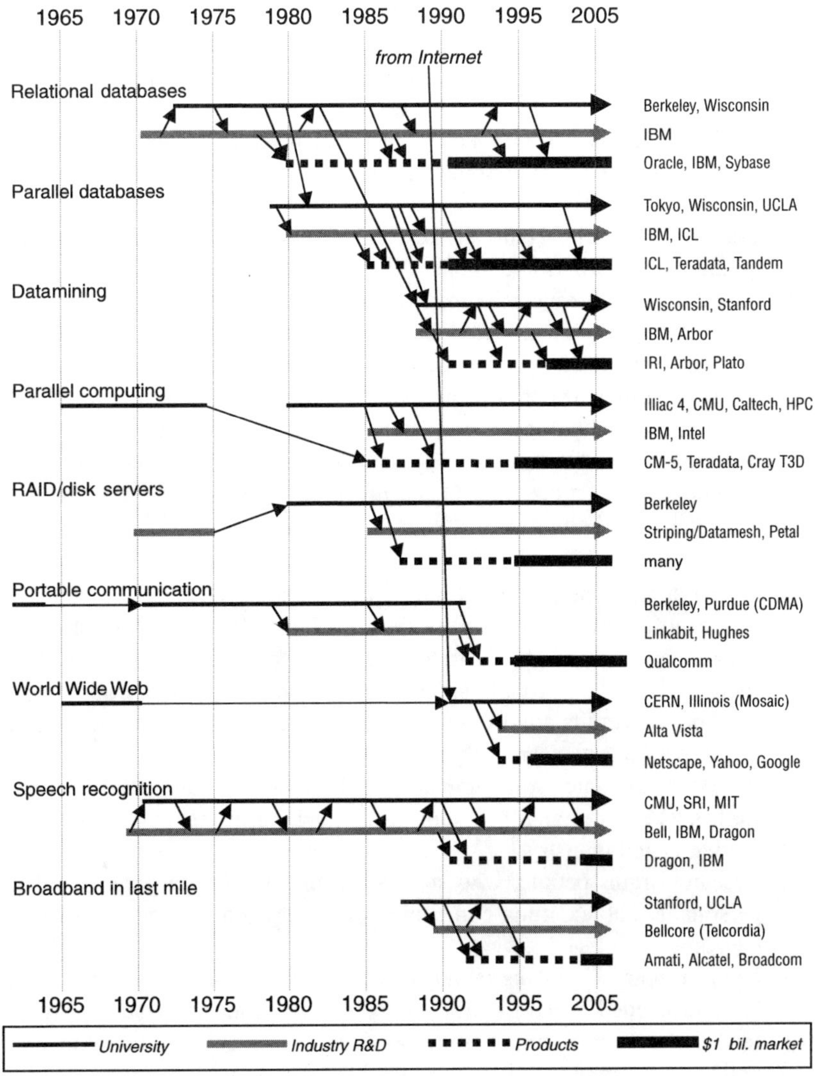
1965 1970 1975 1980 1985 1990 1995 2005
from Internet
Relational databases
Berkeley, Wisconsin
IBM
Oracle, IBM, Sybase
Parallel databases
Tokyo, Wisconsin, UCLA
IBM, ICL
ICL, Teradata, Tandem
Datamining
Wisconsin, Stanford
IBM, Arbor
IRI, Arbor, Plato
Parallel computing
Illiac 4, CMU, Caltech, HPC
IBM, Intel
CM-5, Teradata, Cray T3D
RAID/disk servers
Berkeley
Striping/Datamesh, Petal
many
Portable communication
Berkeley, Purdue (CDMA)
Linkabit, Hughes
Qualcomm
World Wide Web
CERN, Illinois (Mosaic)
Alta Vista
Netscape, Yahoo, Google
Speech recognition
CMU, SRI, MIT
Bell, IBM, Dragon
Dragon, IBM
Broadband in last mile
Stanford, UCLA
Bellcore (Telcordia)
Amati, Alcatel, Broadcom
1965 1970 1975 1980 1985 1990 1995 2005
University
Industry R&D
Products
$1 bil. market

from a research result to advanced development to product development to economic return. Ideas and people tend to bounce around; new ideas can be stymied by political or business impediments and forced to take alternative routes. The industry does not have a few mechanisms for creating and exploiting innovations. Instead, it enjoys what has been called a "national research culture" that fosters innovation (Lazowska, 1998). Some of the features of that culture are described below.

A 1995 report of the National Research Council Computer Science and Telecommunications Board documents the effect of the research culture in the computer and high-performance computing arenas (NRC, 1995). A more recent report (NRC, 2003), documents how technologies are born (often in academia), are taken up and extended by other academic or industrial groups, become the seeds of start-up companies or new products in larger companies, as well as how the market for the technology grows and matures (Figure 2-1). All paths to market are erratic, and often take 15 years. The diversity in the academic and industrial sectors lends robustness to the process: a good idea is very hard to completely eradicate. Among the findings of that report are:

- Research has kept paying off over a long period.
- The payoff from research takes time. At least 10 years, more often 15, elapse between initial research on a major new idea and commercial success. This is still true in spite of today's shorter product cycles.
- Unexpected results are often the most important. Electronic mail and the "windows" interface are only two examples.
- Research stimulates communication and interaction. Ideas flow back and forth between research programs and development efforts and between academia and industry.
- Research trains people, who start companies or form a pool of trained personnel that existing companies can draw on to enter new markets quickly.
- Doing research involves taking risks. Not all public research programs have succeeded or led to clear outcomes even after many years. But the record of accomplishments suggests that government investment in computing and communications research has been very productive.

Mobility of Ideas

As Figure 2-1 suggests, the number and types of research structures among universities and industry provide a kind of redundancy; an idea that cannot advance in one environment may flourish in another. As an example, the path of reduced instruction set computing (RISC) started out with John Cocke's IBM 801, developed at IBM Research. Although the ideas were countercultural and did not have a great impact at IBM, they spawned two research projects, one at Berkeley and one at Stanford, to explore them further. Both university projects resulted in prototype processor designs. The Stanford project formed the nucleus of a start-up company—MIPS Computer—to build RISC microprocessors. The Berkeley project led to an advanced development project at Sun Microsystems to develop its SPARC instruction set. Both led to commercially successful products. Moreover, publication of the work in professional journals rapidly spread awareness of the technology. A related example occurred in the evolution of relational databases. A researcher at IBM, Ted Codd, developed an idea that found little encouragement at IBM, whose products at the time used a competing database technology. Nevertheless, Codd was able to seed academic work at Berkeley that enlarged the interested community and eventually led to two start-up companies, Ingres and Oracle, and a huge industry (NRC, 1999). Both processor and database technologies were later embraced by IBM.

Table 2-2 shows one way that ideas can move between academia and industry. A key idea of this sketch is the "democratization" phase, in which a tiny research community is deliberately enlarged into a community with a critical mass of researchers exchanging ideas, building prototypes, and teaching others. This step was clearly discernible in the histories of both RISC and relational databases, and was, perhaps, the key step in the spread of very large-scale integration design techniques that Carver Mead and Lynn Conway developed in the late 1970s. Democratization in that case involved writing a textbook, teaching teachers, and starting courses in a half-dozen graduate departments to spread the ideas. These efforts resulted in a self-sustaining research community that built computer-aided design software, built a short-run chip fabrication system (MOSIS), designed a number of novel chip architectures, and trained hundreds of students in the art of integrated-circuit design and computer architecture.

TABLE 2-2 How Ideas Can Move between Academia and Industry

University	Industry/Government
1. Theoretical result →	2. Student graduates to industry laboratory that encourages individual researchers and builds a basic prototype.
3. Democratization phase, in which many people work on the idea (e.g., RISC) →	4. Advanced development leads to a commercially successful product in a small but significant market.
5. Academics study the details and fill in the gaps [lots of difficult research here] →	6. Market explodes. Industrial research advances technologies.

Source: Tennenhouse, 1998.

Mobility of People

People often move on to new challenges, sometimes taking with them innovative ideas. The industry asserts, "technology transfer is a contact sport," that ideas transfer best when people carry them. Universities are, of course, a primary source of people, graduate and undergraduate students, many of whom have had research experience. University graduates with research experience are very valuable to industry, not only as staff for its research functions, but also as technical leaders in product development organizations. Curious people with broad technical knowledge who are trained in solving technical problems through research are extremely valuable in today's product engineering groups.

People with ideas often feel impelled to find a receptive environment, and in the 1990s start-up companies were a powerful attraction. University graduates sometimes embarked immediately on a start-up company, based on ideas formulated or prototyped while they were students. Faculty members often took leave to start companies or to consult with companies that embraced their new ideas. In fact, departing faculty members and students left some academic computer science departments with large gaps in their curricula and research programs, especially systems and networks programs (Morris, 1998).[3] The lures are not only financial. Many students who would otherwise go to graduate school complain that academic research is sterile and irrelevant; they prefer to actually *do* engineering, to build a product that will change the world.

The mobility of people includes flow from industry to academia. For instance, in a 2001 survey, many computer science programs reported record numbers of applicants for doctoral programs, attributable to the demise of many Internet start-ups (Bryant and Vardi, 2002). Anecdotal evidence also suggests that many industry researchers have found places in universities as industrial research spending has declined at telecommunications and computer hardware companies. Their knowledge and industrial experience can provide valuable insight for academic research.

Open Structures

The structure of the network systems and communications industry promotes certain kinds of innovation. When components have well defined interfaces, innovators can offer improved components with the same interfaces, so they remain compatible with their predecessors. Thus, there is a ready-made market for the new product. Moreover, some interfaces are specifically designed to accommodate innovation. For example, the "operating system interface" invites application programmers to write innovative applications by eliminating the need to deal with a myriad of details of hardware control. Standards for communications and network protocols (interfaces) allow innovative products to interoperate with or supplant predecessors.

An "open" interface by definition is an interface controlled by a consensus of the interested parties. Formal standards, for example, are open, and the industry has many industry consortia that develop and maintain open standards (e.g., IETF, X/Open, OMG). Open interfaces promote innovation by providing innovators with a dependable, stable interface for new products. The success of the deregulation of telecommunications services depends on standard interfaces. For example, a compatible local exchange carrier must be able to connect into the networks of other local exchanges and long-distance carriers with predictable interfaces.

Software

Software is the universal building material of network systems and communications products and services, and the importance of software as an enabler of innovation cannot be overemphasized. The vital functions of many products and services are controlled by software. Many innovations are merely software improvements, but dramatic innovations sometimes come from relatively simple software. For example, the World Wide Web is essentially a set of common standards applied to a preexisting Internet communications infrastructure enabled by browser software. Another example is an easy change in software to use a digital compression algorithm to increase the effective speed of a modem; this change was based on digital signal processors becoming fast enough to implement a high-speed modem in software alone. An encryption feature could also be added easily.

Software coupled with telecommunications has another virtue—updated software can be distributed rapidly to customers over a network. Whether the new version fixes a bug or introduces a new feature, customers can easily upgrade their equipment, and new software systems are frequently introduced by allowing free downloads over the network (e.g., the Netscape browser, media players). By distributing software widely at zero or low cost, firms count on network effects to generate even broader use and to build a strong market base for their products.[4]

Software can be customized to meet the special needs of individual customers. Although many vendors do not offer variants because of the expense of testing and maintaining separate versions for separate customers, "open source software," makes the source code available to customers allowing them to innovate independently. Although some large pieces of software (e.g., the Linux operating system and Gnu software tools) are available in open source, it is too soon to tell whether open source software will become a significant pattern in the industry.

Multiplier Effect of Infrastructure

Infrastructure is critical to the advancement of network systems and communications technologies. At any given time, the installed networking and computing facilities are the substrate upon which further innovations are developed and introduced. The innovations may, in turn, lead to pressure to enhance the infrastructure, thus initiating a new cycle. The early ARPAnet is an example; the need to connect ARPAnet to other networks led to internetworking protocols, most notably the transmission control protocol and Internet protocol (TCP/IP) (Cerf and Kahn, 1974). The demand for connections led to higher transmission speeds, faster routers, and routing protocols that could be scaled to a larger network. The larger network and its protocols and naming conventions, in turn, provided the near-universal connectivity that led to the creation of the World Wide Web. The growth of the Web increased the demand on the capacity and scale of the network, and today, the infrastructure is being challenged to carry traffic with real-time requirements, such as VoIP (voice over IP) and video.

Advances in network infrastructure have been a key to fostering innovation. The federal government made the initial investments in ARPAnet and NSFnet. As the network expanded to nonacademic customers, regional network consortia built up the network. Today, service revenues support the network, but the federal government continues to support experiments that may lead to significant improvements in performance (Internet2). A similar pattern of infrastructure investment occurred in academic computing facilities. When workstations were first introduced, the National Science Foundation (NSF) helped equip academic research centers with the new technology, which served as a substrate for academic research in networking and interactive computing. Subsequently, parallel computers were provided to encourage research on software tools for writing high-performance parallel computing applications.

Intellectual Property

In addition to patents, the industry also issues many licenses and cross-licenses. No company has a dominant position in the industry based on intellectual property (in contrast to the way Xerox dominated the copier industry when its basic xerography patents were still in force). Patents covering interfaces—whether computer buses or communications protocols—must be licensed widely because interfaces must be open to be widely used. Therefore, to receive their endorsement, most standards bodies require that patents covering standards be licensed liberally.

Size of the Research Investment

Government Funding

The federal government has provided major support for university computing and networking facilities. The Defense Advanced Research Projects Agency

(DARPA) and NSF were particularly active in the development and growth of the Internet and continue to provide the bulk of support for new initiatives in networking research and infrastructure. Federal funding for research in computer science increased from $129 million in 1980 to $1.5 billion in 1999 (NSF, 2001). In 1999, roughly 33 percent of this funding was provided to universities—the rest went to industry and government agencies; more than 75 percent of funding for basic research went to universities. In electrical engineering, of which communications is a subset, federal funding for research remained basically flat throughout the 1980s and 1990s, peaking at $881 million in 1993 and retreating to $699 million in 1999 (NSF, 2001). The share of total funding for electrical engineering research that went to universities rose, however, from 10 percent to 27 percent during this period. The two federal agencies that support research in computing and communications are the U.S. Department of Defense (DOD) and NSF, in that order.

Federal funds support roughly two-thirds of university research in computer science and electrical engineering. Some of this funding is used to support the acquisition of research equipment and to support graduate students. The number of graduate degrees in electrical engineering and computer science increased rapidly through the 1980s and the 1990s; the number of master's degrees awarded more than doubled; 925 doctoral degrees in computer science and 1,596 in electrical engineering were awarded in 1998 (Hill, 2001). The proportion of nonresident aliens in total doctoral degree enrollments in computer science and computer engineering has risen steadily since 1945, up to 55 percent in 2001. Interestingly, data from the most recent Taulbee Survey indicate that only 17 percent of new faculty are nonresident aliens; proportionately fewer foreign students take positions at U.S. universities (Bryant and Vardi, 2002).

Industry Funding

Computer-related industries tend to be R&D intensive. Firms in this sector spend a greater percentage of sales revenues on R&D than any other industry except medical devices and pharmaceuticals. In computer-related industries, roughly 10 to 20 percent of corporate R&D funds are spent on research (rather than development). According to a 1999 report by the National Research Council, "Such expenditures tend to derive from, and result in, the fast pace of innovation characteristic of the field" (NRC, 1999). Although the volume of R&D investment in computer-related industries has kept pace with the growth of business over the past decade, the R&D spending of the telecommunications component of the network systems and communications sector has contracted in the wake of AT&T's divestiture, deregulation, and most recently, deep recession in the telecommunications industry.

Although the amount of industry support for university research in network systems and communications is not known, overall industry support for research in science and engineering in universities represents about 7 percent of the total universities receive. The percentage is higher, perhaps as high as 15 to 20 percent, in engineering; support varies by the rank and reputation of the university program (Morgan and Strickland, 2000). For the most part, computer-related industries have tended to draw on academic research more extensively than the telecommunications industry.

Some of the largest firms in information technology provide significant support for university research. IBM currently spends several hundred million dollars per year for research at universities. Support for university research provides IBM with access to activities at universities and contact with potential future employees. Microsoft has established a large research organization that emphasizes fundamental research. It too has established research collaboration with a number of universities, including Southern California, Utah, Yale, and West Virginia University. Other firms have established relationships with a small number of universities. One example is the AT&T Center for Internet Research at UC-Berkeley that was funded by AT&T in 1998 for three years. Another example is Intel Corporation, which has established research sites at the University of California at Berkeley, Carnegie Mellon University, and the University of Washington among others (*http://www.intel-research.net/*).

Some research organizations require or encourage both industry and government funding. NSF supports university-based engineering research centers and science and technology centers that must have industry contributions to supplement government funding. Initiatives like Internet2 and Next Generation Internet, which are funded principally by government, solicit industry support.

CONTRIBUTIONS OF ACADEMIC RESEARCH

Academic research has made essential contributions to the network systems and communications industry. The special strengths of universities are reflected in the ways they have contributed to the industry:

- **Human capital.** Undergraduates and graduate students educated in universities have become key players in industry as individual researchers, development engineers, technical leaders, and entrepreneurs. Research experience in universities is highly valued by industry even for nonresearch employees. As students and faculty flow to industry and start-up companies, they provide an effective form of "technology transfer."
- **Long-term fundamental research.** With proper funding, academic research is able to work on long-term problems that may be ignored by industry or may even be anathema to dominant industry businesses, technologies, regulations, or prejudices.
- **Intellectual diversity.** Academia provides an open setting that can engage colleagues in various disciplines and industries; the results are

reported in the open literature. Concurrent research projects and different approaches provide a kind of redundancy and expand the community of researchers on promising topics. Shared artifacts of experimental research, especially software, are an important way to disseminate research results.

- **Collaboration with industry.** Direct collaboration between industry and academia, both on specific projects and in longer term relationships, has produced significant contributions to network systems and communications. There are many collaborative structures but no dominant or "best" collaboration scheme.
- **Test beds.** University laboratories can serve as test beds for new technologies. Most of the early participants in the ARPAnet, for example, were universities, which played an important role in testing and refining the technology. The pattern has continued with the Gigabit Testbed, vBNS, and other networks, such as a campus-wide wireless network at Carnegie Mellon University.
- **Nuclei for start-up companies.** University research can lead to technologies and people that become the seeds of new businesses. Examples are Google and Yahoo, both spin-offs of research at Stanford University.

At an October 1998 NAE workshop to collect information and exchange ideas for this study, the participants came to the following conclusions: (1) the network systems and communications sector has benefited greatly from a national research culture in which individuals move frequently between academia and industry, thereby increasing their knowledge of both and their contributions to both; (2) personal relationships are crucial; and (3) universities not only invigorate the research culture with fresh students each year, but they also house open research projects that anchor technical disciplines.

Human Capital

Industry looks to universities to educate and train students who will staff industry R&D projects. Industry considers human capital to be the most important product of universities—even more important than new knowledge captured in research results. The question of whether industry wants students with a broad technical education or with training in specific skills, such as programming in a given computer language or the operation of a certain kind of computer or communication device, is answered differently by different businesses. Larger companies tend to prefer broadly educated candidates who can learn skills quickly on the job. Smaller companies that do not have people to serve as mentors and trainers prefer trained candidates.

Training in research is extremely important to innovation, even if an individual does not continue to do research. Industry considers research experience valuable because it demonstrates the abilities necessary for any technical endeavor: self-motivation, problem solving, teamwork, knowledge of related research, contact with other researchers and colleagues, the ability to organize an amorphous problem, and the perseverance to overcome setbacks. Graduates with advanced degrees have already shown greater than average ability; and research training is considered evidence that an individual can tackle difficult technical problems, such as designing and building complex systems.

Students of electrical engineering and computer science have typically been in great demand, not only by companies in the network systems and communications sector, but also by other companies trying to modernize their information technology. The Bureau of Labor Statistics estimates that almost 4 million information-technology workers will be needed by 2010, compared with 1.9 million in 2000 (Bureau of Labor Statistics, 2001). If this projection remains accurate, the current rate of graduation in U.S. universities—approximately 27,600 undergraduates and 900 Ph.D.s in computer science per year—will not bridge the gap (Hill, 2001). Students and faculty who participate in start-up companies are important to the culture of innovation. A significant number of network systems and communications businesses have been founded by students straight from universities or by faculty, who either take a leave of absence or leave permanently. These companies are often formed to exploit a technology developed in the university. One of the best-known examples is Sun Microsystems, which began as a start-up company to commercialize a computer workstation designed at Stanford and Unix software originally conceived and developed at Bell Laboratories by Thompson and Ritchie then further developed by Bill Joy at UC-Berkeley. In the summer of 1998, six (out of 60) members of the electrical engineering faculty were on leave from Stanford University to work with start-up companies. Faculty members who return to the university report that their research has been stimulated greatly by their experience. A founder of Granite Systems, for example, said that he now has a far better sense of what it takes to produce a product, as well as the state of the industry (D. Cheriton, Stanford University, personal communication, September 8, 1998).

A report by BankBoston on the impact of a research university concluded with the following statement: "If the companies founded by MIT graduates and faculty formed an independent nation, the revenues produced by the companies would make that nation the 24th largest economy in the world. The 4,000 MIT-related companies employ 1.1 million people and have annual world sales of $232 billion" (BankBoston, 1997). MIT-related firms are especially prominent in electronics and software. More than half of the companies founded by MIT graduates were founded by graduates in electrical engineering (which at MIT includes computer science).

Research in Engineering and Computer Science

University research in electrical engineering and computer science has made significant contributions to the network systems and communications industry. In some cases, academic research projects have been essential to the creation of billion-dollar businesses (see Figure 2-1 for some examples).

Academic research has also built a foundation of techniques and analysis tools that are widely used as enabling tools by the industry. These include techniques for the optimization of computer programs, for the automatic design of integrated-circuit chips, and for the verification of bus specifications. These techniques are not dramatic developments that spawned businesses, but they have been important to the industry as a whole. The digital cryptographic techniques widely used today to ensure privacy and authentication in electronic commerce and related applications were invented in academia (NRC, 1996). The development of object-oriented programming, from the first step (Simula 67) to the most recent form (Java), took 30 years. Most of the research was necessarily conducted in academia, because industry typically does not invest in risky research that offers only long-term prospects for payoff.

The two case studies below illustrate how academic research has contributed to the network systems and communications sector: (1) the Internet, which shows a 30-year trajectory of academic and industrial R&D to build a revolutionary communications technology; and (2) research in signal processing that led to a start-up company, Amati Communications, that successfully exploited the technology.

Case Study: The Internet

Academic research played a key role in the development of internetworking, the connection of disparate networks into a worldwide, scalable, packet-switched network. The Internet, which now connects more than 100 million people and computers, was the direct result of government-funded research begun in the late 1960s to link different kinds of academic research computers. Although industry was essential to the scaling of the Internet and the development of services, the early technical development depended almost entirely on university research (for a more detailed case study, see SRI International, 1997; for a brief chronology, see Box 2-2). The ARPA funded research, development, and deployment of this revolutionary packet-switching technology, because the telecommunications industry showed no interest in participating.

The story begins with the ARPAnet, which was initiated by DARPA in the late 1960s as a way to share access to expensive or special-purpose research computers around the country. Precursor ideas for packet-switching networks had been developed at MIT and UCLA, but Paul Baran is credited with discovering packet switching while at the RAND Corporation in the early 1960s. Donald Davies, a researcher at the U.K.'s National Physical Laboratory independently

BOX 2-2
Chronology of the Internet Development

1969 DARPA commissions ARPAnet to promote networking research.
1974 Vinton Cerf and Robert Kahn publish a paper specifying the TCP/IP protocol for data networks.
1981 NSF provides seed money for CSnet (Computer Science NETwork) to connect U.S. computer science departments.
1982 DARPA establishes the TCP/IP protocol as standard.
1984 The number of hosts (computers) connected to the Internet exceeds 1,000.
1986 NSFnet and five NSF-funded supercomputer centers are created. NSFnet backbone operates at 56 kb/s.
1989 Number of hosts exceeds 100,000.
1991 NSF lifts restrictions on commercial use of the Internet. World Wide Web software is released by CERN, the European Laboratory for Particle Physics.
1993 Mosaic browser developed at NSF-funded National Center for Supercomputer Applications at the University of Illinois is released.
1995 U.S. Internet traffic is carried by commercial Internet service providers.
1996 Number of Internet hosts reaches 12.8 million.

Source: SRI International, 1997.

discovered the idea of packet switching in 1965, deciding upon some of the same parameters for his network design as Baran, such as a packet size of 1024 bits. ARPA built an early network by contracting with Bolt, Beranek, and Newman to build packet switches (IMPs) at research computers at about a dozen universities. In addition, academic research projects were initiated to develop protocols by which different types of computers could communicate, to outfit the computers with suitable hardware and software interfaces to the network, and to measure the performance of the operating network. Similar networks were also developed, such as a network using satellite or radio-transmission links to connect the packet switches.

Early protocol experiments, together with the clear need for interconnecting the various kinds of networks being developed, pointed to a need for "internetworking." The key idea is the Internet datagram, a universal way of formatting network packets, together with associated protocols (TCP/IP), introduced in a paper by Cerf and Kahn (1974) while Cerf was a member of the Stanford faculty. This paper provided the first definition of Internet architecture and led to implementations and experiments at several universities. With the TCP/IP implementation developed in "Berkeley Unix" software (at the University of California at Berkeley) and released in the late 1970s, it was easy to connect academic research computers to the network. Ad hoc committees of academic researchers refined the TCP/IP protocol standards, including application protocols.

In the late 1970s, academic computer science research centers not served by the ARPAnet banded together to form CSnet, also using the TCP/IP protocol. In

1983, the new network was linked to the ARPAnet, an event that could be called the birth of the Internet. Subsequently, the network continued to grow, and the demand for connections increased. In 1986, with the creation of NSFnet, the responsibility for the principal "backbone" of the nationwide network shifted from DARPA to NSF. The network could be used only for research and education, and academics continued to play a major role in network governance and engineering.

With the emergence of the World Wide Web, the Internet was no longer only for research and education but became a worldwide network connecting businesses and consumers, as well as researchers. The idea of browsing text documents obtained in a uniform way from any machine connected to the Internet was developed in 1991 by Tim Berners-Lee, then at the CERN nuclear research facility in Geneva, Switzerland. Documents on the Web may contain "hyperlinks" to other documents, thus linking documents into a complex "web." Marc Andreesen and other researchers at the University of Illinois later extended the Web to include pictures and other types of data. They also built a graphically oriented browser, called Mosaic, that allows users to "click" to follow a hyperlink, thus opening browsing to a wide range of people. Jim Clark recruited Andreesen to cofound Netscape Communications, which developed Netscape's Navigator browser product based on Mosaic. Microsoft soon developed a competing product, Internet Explorer. The combination of pictures, ease of use, and supported products enabled the Web to grow with astonishing speed. It quickly developed into a mechanism for publishing, for finding information, and for transacting business electronically.

As more computers were connected to the network, the bandwidth and switching capacity had to be expanded. DARPA and NSF, with university and industry support, organized a series of test beds to explore high-speed networking technologies and test emerging products and protocols. Between 1990 and 1994, NSF and DARPA funded the Gigabit Testbed Initiative, a university-industry-government effort to explore networking technologies at speeds of 155 Mb/s and higher; one of these test beds achieved long-haul transmission at 800 Mb/s. NSF operated the vBNS network (very high-speed backbone service) in conjunction with MCI to link more than 75 universities in a network with backbone speeds of 622 Mb/s to 2.4 Gb/s and access links of 43 Mb/s to 155 Mb/s until the vBNS network was terminated in April 2000. The participants explored new applications of advanced communication bandwidth and protocols, as well as operational and governance issues. Universities are presently engaged in a new round of infrastructure enhancement, Internet2, designed to meet a full range of academic research needs.

As the Internet expanded, commercial businesses and services grew up alongside government and academic operations. New companies were started to sell packet switches (routers), application software, authoring tools, and network services. The leaders were not the telecommunication companies, but start-up companies, such as Cisco and Netscape, using technologies developed in universities. Network service providers emerged as regional networks (e.g., NEARnet, BARRnet) and were encouraged to connect to the national backbone operated as part of NSFnet. International connections were also developed. As the Internet grew, major telecommunications companies began to offer Internet service as well. In the United States today, businesses and residents in most densely populated areas have a choice of several Internet connection methods and service providers.

The governance and engineering of the Internet are unique: a governing body (the Internet Activities Board) and engineering standards organization (Internet Engineering Task Force) consist of volunteers from Internet participants. At first, committees of academics and the few government contractors that were building and operating the network set protocol and other engineering standards. Today, these committees have broad participation from academia, industry, and nonprofit organizations.

Academic Contributions

An SRI study commissioned to analyze the nature of the research that contributed to the Internet described the contribution of academic research in some detail (SRI International, 1997):

> The Internet appears, overall, to be primarily a problem-driven, technology-based innovation that required little *direct* input from fundamental research for its realization. The driving forces, interestingly, were not profit incentives in the private market, but public goods, first in the realm of national defense and subsequently in the university and government research infrastructure, as a means of fostering communication among computer scientists. What we are calling the Internet's intrinsic technologies—network design, packet switching, routers, protocols, browsers—were the products of problem-driven research conducted in universities and government contractor laboratories with government support. One possible exception is the research conducted at the University of Illinois' NCSA [National Center for Supercomputer Applications], which took place in an environment (according to Andreesen) that enabled researchers to head off in directions that looked "interesting" without seeking justification. Nonetheless, the context was one of *application,* as suggested by the Center's name. Although the evolution of the Internet did not encounter technical roadblocks that required fundamental research for their resolution before further progress could be made, there is obvious, fundamental research content in both the Internet's intrinsic and supporting technologies. The electronic and physical infrastructures that comprise the Internet clearly depend on information theory, solid-state physics, electro-optics, and other fields on which modern communications technology is based and for which NSF has provided substantial support.

The SRI study stresses the importance of government, industry, and universities in the development of the Internet and points out that, as the focus of technical and organizational innovation shifted from government to industry over a 30-year period, universities played a constant supportive role. By following the career trajectories of some key individuals who moved among the three sectors, the SRI study highlights the importance of networks of individuals and the importance of human capital. The study also cites the frequent opportunities for interactions and linkages among the three sectors and the ease with which individuals can move across sectors in the United States as important factors to success (SRI International, 1997).

The Internet is an example of the major impact academic research can have in the creation of an entirely new technology. Although the enormous impact of the Internet is unusual, this case illustrates how government, academia, and industry can contribute to technological—and therefore economic—success.

Case Study: Amati Communications

Amati Communications was founded in 1991 by Stanford University professor John M. Cioffi to commercialize technology for transmitting high-speed digital signals over copper telephone wires. The technology, named discrete multitone (DMT) modulation, was one of several technologies competing to be adopted as an industry standard. DMT is now an accepted standard for providing DSL (digital subscriber line) service and a commercial success.

Research

The original work on DMT was conducted in 1987 by a research group at Stanford directed by Cioffi, who was then an assistant professor. Cioffi used funds from an NSF Presidential Investigator Award (1987–1992), with matching funds from several companies, including Bell Communications Research, to investigate asymmetric digital subscriber lines. The initial objective was to develop reliable transmission of high-quality digital movies over phone lines, which required speeds about 10 times faster than integrated service digital network (ISDN) lines, the existing technology. Later, the objective evolved to encompass high-speed Internet access and other data applications. The researchers investigated many methods and focused on an old encoding technique called multitone transmission, in which separate frequency channels (tones) carry separate digital signals. A crude analogy would be sending several channels of Morse code over a telephone line, with each channel using a different audible frequency (like the seven separate tones used in touch-tone dialing). A receiver can split out the separate tone frequencies and decode each Morse sequence.

The goal of Cioffi's research was to transmit data as fast as possible, which would require using many separate frequency bands and sophisticated signaling techniques (not Morse code!) in each band. This objective led Cioffi and his team to seek fundamental improvements in digital signal processing algorithms that could be applied to various channels. But the most important innovation was the adaptation of each band to the band's transmission characteristics. In effect, DMT measures the properties of signals transmitted over each band and then allocates data accordingly. A band that attenuates signals less than another band carries more data. A band that introduces less noise than another also carries more data. DMT also measures and compensates for the transmission characteristics of each pair of copper wires. This complex scheme is practical because of inexpensive digital signal-processing hardware.

Some of the key innovations for DMT were patented while Cioffi's group worked to refine, promote, and publish the method in IEEE journals and at American National Standards Institute standards meetings. Three of the patents, which are assigned to Stanford, are still valuable and are licensed throughout the telecommunications and data communications industry. At least one patent, on pioneering artificial intelligence techniques used in the adaptation, is considered necessary to comply with any of the existing or emerging DMT standards. In addition to patents, Cioffi and his graduate students acquired valuable know-how that would benefit any company that attempted to use DMT.

Amati Start-up Company

In 1991, after an unsuccessful search for corporate partners, Cioffi founded Amati Communications Corporation with Stanford University and a Stanford alumnus, Mr. Kim Maxwell, who was the company's first CEO. Amati's vision was to "get DMT on every phone line in the world." To be successful, Amati had to make DMT an industry standard, which involved working with national and international standards organizations and competing with other technologies. On March 10, 1993, after competitive testing against several other technologies, including an alternative promoted by AT&T, DMT technology was selected unanimously as a U.S. standard.

As part of Stanford's contribution to the company, Cioffi was given three years of leave (with 50 percent leave spread over two years to make up the third year). In exchange for an exclusive licensing and sublicensing privilege on the DMT patents, Stanford received stock and a promise of royalties. In 1997–1998, Stanford received $7.9 million when it liquidated its holdings in Amati; royalties totaled more than $8 million in 2001 (Stanford University Corporate Guide, 2001).

Between 1991 and 1998, Amati employed several of Cioffi's graduate students, at least four of whom were directly involved in the Stanford research and had considerable knowledge of DMT. Stanford's then dean of engineering, Jim Gibbons, was chairman of Amati's Board of Directors until 1998.

Amati went public late in 1995 and was a growth leader on the NASDAQ for approximately one year. In March 1998, Texas Instruments acquired Amati and the DMT license for approximately $450 million in cash. At the end of 2001, 3.6 million residential DSL (digital subscriber lines) using DMT technology were installed in the United States, which was then projected to grow to 13.5 million by 2005 (InternetWeek, 2001).

Academic Research in Economics, Social Sciences, Management, Design, and Policy

Academic research in a variety of nonengineering disciplines has also contributed to the network systems and communications sector. This research has focused on how computers and communications systems fit into larger socioeconomic systems. Examples include how information technology systems increase business productivity; the effects of e-mail on people and organizations; how people use the Internet; the effect of the Internet on family structures; the effect of prices on communications services; and the value of communication services to consumers.

Some of these studies have formulated and tested models to explain the behavior of people and systems in operation. For example, the GOMS model of performance grounded in cognitive psychology (Card et al., 1983) has been used to design interactive interfaces. John Anderson has built several successful computerized tutoring systems based on a detailed understanding of how people model specific subjects (e.g., geometry and algebra) and the errors in these models. Results of social psychology research on electronic communication have been used to improve training for new users, in effect teaching them about the social norms and effects that spring from the technology.

In some cases, the models only described what had been observed, but in a few cases they were used to predict the behavior of future systems. For example, models can predict how well people will perform simple interactive computer tasks. Although models cannot predict whether one chat room or e-mail system will be more popular than another or the details of e-mail usage, general principles can be learned. For example, people are generally less inhibited when using an electronic communication medium than in face-to-face interactions. Observations such as this can lead to a better understanding of network systems and communications, but they cannot be used as a guide to design. At a workshop held in connection with this study, the consensus was that these research topics should be given more attention, especially as the services provided by network systems and communications become more important and affect a greater portion of society.

Economics, Policy, and Regulation

Because communication systems have historically been operated as regulated monopolies, researchers in economics and policy were able to study them extensively. The work of one academic, Alfred Kahn of Cornell University, was used as a basis for the deregulation of several industries, including trucking, airlines, and communications (Kahn, 1970, 1971). Nevertheless, a great many questions about economics and policy remain to be answered (see Box 2-3).

BOX 2-3
Contributions of Economics and Other Social Science Research to the Development of Information Technology

Role of regulation. Economics research by Alfred Kahn (1970, 1971), Paul Joskow, Roger Noll, and Kip Viscusi redefines the role of regulation from protecting the public interest to stepping in when markets fail to drive prices to marginal costs. This redefinition has helped spur deregulation in a number of industries, including communications.

Network externalities. Work by Hal Varian, Paul David, Brian Arthur, Garth Saloner, David Shapiro, and others shows that the network industries and information industries are characterized by "network externalities" that make the value to a consumer of a particular product or service increase as more people use it. An example is an Internet connection that becomes more valuable with the amount of information available and the number of people connected. This insight reinforces the importance of getting products and services to the marketplace quickly, pricing them low at first to establish markets, and then raising prices as more units are sold and their value grows.

Internet Economics. Research by McKnight and Bailey (1997) and others address the implications of the pricing of Internet-based resources and services, such as the allocation of resources based on the willingness of users to pay.

Group dynamics and decision making. Research by Sara Kiesler, Suzanne Iaconno, Wanda Orlikowski, and others (e.g., Siegel et al., 1986) on group dynamics and decision making in small electronic groups informs the design of group decision-support systems.

Diffusion of applications. Research by M. Lynne Markus (1987) and others examine how critical mass predicts the diffusion of networked applications within organizations and informs the deployment decisions for information technology applications.

Distribution of the benefits of information technology. Research by Tora Bikson, Lee Sproull, and others demonstrates that peripheral members of social systems benefit more from using electronic communication than central members (e.g., Sproull and Kiesler, 1991) influencing policy decisions about subsidies for access to the Internet.

Information sharing. Research by Paul Attewell, Tora Bikson, Sara Kiesler, Robert Kraut, Lee Sproull, and others demonstrates how personal attributes and organizational characteristics such as incentive systems influence peoples' use of information technology for information sharing. Research by Julian Orr (1990) and others demonstrates that service technicians often have more useful technical expertise than system designers and share their knowledge in a community of practice. This work influenced the design of a community-based troubleshooting database at Xerox Corporation that has significantly improved service performance (Bell et al., 1997).

Sources: Sirbu, 1998; NRC, 2000.

The Telecommunications Policy Research Conference is an annual forum for scholars engaged in research on policy-relevant telecommunications and information issues and public-sector and private-sector decision makers engaged in telecommunications and information policy. The wide range of topics at the 2002 conference reflected the intense academic interest in telecommunications policy. Topics included: comparative telecommunications policies in the United States and abroad; broadband deployment and uptake; spectrum management; computer and Internet security; wireless communications standards; mergers and acquisition; intellectual property; basic research in telecommunications; mass media; and numerous other topics. University researchers presented the majority of research results at these sessions, reflecting the active involvement of academic research.

Business and Management

Business schools have long been concerned with how information technology can be exploited for the benefit of businesses (see Box 2-4). For example, research on decision-support systems not only developed techniques for collecting and presenting relevant business data to management, but also compared the quality of decisions made with different kinds of supporting technology. The rapid development of the Internet has opened up new avenues for study, such as supply-chain integration, which uses the network to connect a manufacturer's process-planning system to the corresponding systems of its suppliers to ensure a smooth flow of the component parts required to fulfill orders. Success will depend on solving problems related to information technology, network protocols, and control theory. Electronic commerce will certainly face new problems that must be addressed.

Optimizing network design in network systems and communications businesses is a similar problem to the transportation problems studied by operations research. In the early development of the ARPAnet, attention was focused on optimization of network design. Today, the emphasis is on techniques to expand networks to meet burgeoning demand.

Psychology and Social Sciences

Research by psychologists and social scientists has focused on how people use computer and communication systems, the effects of these systems on people, how people interact with each other, and how they work in organizations. These studies are retrospective, conducted after the technology has been deployed long enough for transient behaviors to abate. De Sola Pool's classic book, *The Social Impact of the Telephone* (1997), is a fine example. Other examples are Computers in Classroom Culture (Schofield, 1995), *Connections* (Sproull and Kiesler, 1991), *The HomeNet Field Trial of Residential Internet Services* (Kraut et al., 1996), and *The Second Self* (Turkle, 1984), a study of personal interactions with computers.

BOX 2-4
Contributions of Business Research to the Development of Information Technologies

Critical success factors. Rockart (1981) identifies factors critical to the success of information systems in business settings. The author addresses the question of which information is critical to the success of a business; the questions a database should answer; and how information systems can be designed to support business objectives.

Decision-support systems. Work by Keen and Scott Morton (1978) promulgates the idea of using information systems to support corporate decision making at a variety of levels.

Information technology and strategic management. Research by Earl (1988) stresses that IT is not just for back-office operations but contributes to a firm's competitive advantage. Companies that deploy and employ information technology systems wisely can benefit in the marketplace.

Computer-supported cooperative work. This research introduces the idea of using information technology to allow people to work cooperatively within and across organizations, thereby overcoming differences in geography or time.

Productivity and information technology. Productivity gains from investments in information technology have been hard to measure, but Brynjolfson's (1991) analysis of firm-level data (as opposed to industry-level data) indicates that investments may have large payoffs, but not immediately. The author identifies factors that contribute to positive returns from investments and the characteristics of firms that do and do not experience increased productivity.

Software development methodologies. Research by Cusumano (1991) provides guidance on software development methodologies from the point of view of management.

Process handbook. This repository of business-process knowledge developed by the MIT Process Handbook Project (Malone et al., 1999) can facilitate further research and help determine best practices for deploying information technology. The classification and structure of the database itself is a powerful tool.

Source: Based on Malone, 1998.

Research has also helped guide the design of computer and communication systems. *The Psychology of Human-Computer Interaction*, a classic work by Card, Moran, and Newell (1983), showed how studies in cognitive psychology could be used to estimate human performance when interacting with a computer. These and other performance studies have influenced the design of graphical user interfaces. Ethnographic studies of the behavior of boys and girls at play have been used to inform the designs of many products.

Design Research

Several universities have developed broad multidisciplinary programs aimed at harnessing developments in information technology to human needs. A leader in this area, the MIT Media Laboratory, brings together individuals from a broad spectrum of disciplines, including the humanities and fine arts, to conduct research and application development. For example, the News in the Future Project explored innovative ways to present the news to people using electronic media by tailoring content, presentation, and structure to the needs of the viewer. In addition to developing prototype applications, the laboratory often works on fundamental technologies, such as video compression or image understanding.

MECHANISMS FOR UNIVERSITY-INDUSTRY COOPERATION

Collaborations between industry and academia raise some obvious questions about the kinds of organizations and mechanisms that work best. A questionnaire on the subject sent to 60 researchers for this study elicited a wide range of responses. For example, one respondent felt that "centers which promote close interactions between academic researchers and knowledgeable industrial sponsors are probably a prerequisite for making progress." Another mentioned several collaborative arrangements: joint research programs, like MIT's Project Athena; experimental test beds and university centers, like the NSF-supported supercomputing centers; and consortia. Another respondent felt that "Centers have an indifferent record in communications . . . I doubt that such forms of collaboration will ever be a success." Still others felt that the structure of the organization didn't make much difference as long as the participants understood each other's value systems.

NSF has several programs to create university-based, industry-university research centers and engineering research centers (ERCs), both of which require industrial participation. The ERCs, which are designed to integrate research and education, have generally received favorable reviews (Parker, 1997). However, the Telecommunications Research Center at Columbia University, funded by NSF from 1985 through 1995, was the only ERC established in the network systems and communications area.

The network systems and communications sector does not have an institution comparable to the Semiconductor Research Corporation, an organization that provides industrial support for university research relevant to semiconductors, based on a 10-year technology "road map" to help guide research funding decisions (Bailey et al., 1998). Although the establishment of a consortium of network systems and communications businesses has been discussed, nothing has come of it so far. A consortium of computer storage peripheral companies, National Storage Industry Consortium, has been established to support academic research through focused programs like optical storage. In addition, some firms have targeted their research support for a small number of universities; for example, in late 1998, AT&T and the International Computer Science Institute at UC-Berkeley announced formation of the AT&T Center for Internet Research (ACIR), a multimillion-dollar research center that AT&T agreed to fund for at least three years (AT&T and International Computer Science Institute, 1998). Recently, Intel has sited research operations at UC-Berkeley, Carnegie Mellon University, and University of Washington, all centers where researchers are Intel employees and university professors are engaged as laboratory directors and technical leaders (Intel, 2003).

The Microelectronics Innovation and Computer Research Opportunities (MICRO) Program in the University of California system is an example of a state government effort to encourage university-industry cooperation. The MICRO Program was established in 1981 by the state of California to support innovative research in microelectronics technology, its applications in computer and information sciences, and its necessary antecedents in other physical science disciplines. The program is a partnership between industry and the state in which the state supplements industry funds and waives overhead on university research funding. In 2001–2002, 96 companies contributed approximately $6 million in cash and equipment to fund 98 different projects (MICRO, 2002). In some cases, MICRO support has led to increased federal funding, as well as long-term partnerships between universities and industry. For instance, after an initial concept phase, the RAID (redundant arrays of inexpensive disks) project at UC-Berkeley received MICRO support, which led to the creation of an industrial consortium in 1988–1990. The federal government became a research sponsor in 1990. By 1996, RAID was a $10 billion industry.

In general, university-industry collaborative arrangements in network systems have received mixed reviews. No structure has emerged as the "best," nor has any scheme emerged that works robustly in different circumstances. It appears that strong personal leadership and a collaborative spirit between an academic researcher and his or her industrial counterpart are the elements essential to success. The problem, of course, is that a good collaborative project can founder if one key individual (e.g., the "champion" in the firm) is transferred or moves elsewhere.

FINDINGS AND RECOMMENDATIONS

Academic research has made essential contributions to the network systems and communications sector. Although these contributions—trained researchers, new technologies, algorithms, and prototype systems; early operating experience; studies of social and economic effects—cannot be quantified, they have undoubtedly had a substantial impact. In this industry, the academic ivory tower has been heavily populated by entrepreneurial engineer-researchers.

Findings

Finding 2-1. Academic research has played and will continue to play an important role in the research culture of the network systems and communications industry.

University-industry collaborations are fostered by a vigorous research culture, and academic research has been crucial to the technical evolution of the industry, especially in the development and deployment of the Internet. To be sure, the recent deep recession in the telecommunications sector, which has forced significant reductions in corporate R&D budgets and manpower, has further diminished the industrial research contributions of this important subsector of the network systems and communications industry—a trend begun with the breakup of the Bell system and further deregulation. Given the historical reliance of the telecommunications sector on internal industrial research, changes may be needed. If trends in industrial research persist, academic research in telecommunications will have to be increased. However, for the most part, the research culture that supports the network systems and communications industry is functioning well and needs no major repairs.

People are the key components of this research culture. Collaboration between universities and industry often depend on sometimes fragile personal relationships that can be threatened if an industry researcher is reassigned or an academic researcher goes off in a new direction. Students, faculty, postdoctoral students, researchers with long-term visions, and researchers who focus on applied problems play different roles. Contrary to popular opinion, university-industry projects are not devoted exclusively to long-term basic research; teams of faculty and students often address pragmatic, applied problems in close cooperation with industry.

The flow of people from academia to industry and vice versa is essential to the well-being of the industry and to academic research. The university's role of fueling the research culture with trained students is unique, and training in research is extremely important for innovation, even if the researcher does not continue to perform research but becomes part of an academic-like cadre that pursues innovations in industry (such as the groups awarded the Association for Computing Machinery's Software System Award).

Universities also have a very broad research culture, and network systems and communications systems have increasingly drawn on the wide range of technologies and expertise available at research universities. Electrical engineering and computer science are, of course, central to the industry, but other areas, such as cognitive science, social science, economics, and business modeling, are becoming increasingly important, especially as the importance of information technology-delivered services increases. Some research universities (e.g., UC-Berkeley, University of Michigan, Indiana University) have created information-centered schools that focus on the social, political, and organizational context of information.

To participate in the research culture, a company must have a capacity to absorb innovation, an industrial research laboratory, for example, that can absorb people and ideas from outside the firm and exploit them within the firm. Despite the trend in industry research toward applied problems, industry laboratories have so far retained this absorptive capacity. In the last 15 years, innovations in the industry have focused on completing the digital revolution in telecommunications (e.g., new switching gear); new transmission protocols (e.g., SONET and Asynchronous Transfer Mode), and transmission equipment; faster modems; the refinement and deployment of fiber optics; the refinement of IP protocols; and the widespread deployment of the Internet.

The focus is now shifting toward innovative services, which requires an understanding of psychology, consumer behavior, social phenomena, and other disciplines that can inform the design and operation of new services. The explosive growth of the World Wide Web can be attributed to its social properties more than to its technical capabilities. Industry is most likely to devise and launch new services, but formal research on the uses and effects of these services is most likely to be undertaken in academia.

Finding 2-2. Innovation cycles in the network systems and communications industry have worked well.

An astonishing number of incremental changes have cumulatively taken on the character of breakthroughs. The Internet, for example, had its roots in the ARPAnet in 1969, was developed and deployed incrementally, and was launched into the public arena by the World Wide Web and browsers in 1993. The effect on the industry was of a breakthrough; the telecommunications industry today is utterly different from the industry of 10 years ago. Similarly, the incremental evolution of technologies (e.g., batteries, low-power circuits, integrated radio-frequency elements) to support small portable devices spawned breakthrough products, such as pagers, cellular telephones, and packet-radio modems.

Innovations in the network systems and communications industry have been characterized by the integration of a wide range of technologies: chip designs with increasing levels of integration; digital-analog integration in wireless and wireline communications (e.g., cellular and satellite telephony, wireless devices, such as pagers and security devices, modems and cable modems, local-area network and intranetwork systems and communications receivers, optical network interfaces); internetworking technologies; and, above all, the increasing use of software technologies of all kinds. The convergence of computing and telecommunications has brought together a wide array of technologies for new products.

Academic contributions to these innovations have varied widely. Some innovations originated in university research and were spun off into venture-capital-backed start-up companies. This route is supported by established university policies, a workforce eager to engage in risky start-ups, and a mature venture-capital industry that has been willing to back telecommunications and networking businesses. Many academics have been consultants for network systems and communications companies. Academic design departments have worked on industry projects, and some companies have supported academic research, often directed toward solving specific problems. Sometimes, new companies or products have emerged from business-school entrepreneurship programs.

Finding 2-3. The success of industry-academic collaboration (as defined by participants) depends more on leadership and people than on the type of collaborative structure.

Organizational structures, such as research centers that foster university-industry collaboration, receive mixed reviews in network systems and communications. Success appears to depend less on the choice of structure, the funding arrangements, or the legal agreements than on the leadership and passion of the people involved. A committed leader is essential for establishing personal relationships, and, in general, researchers consider inducements to individuals (as opposed to institutions) more effective than collaborative organizational structures. Dependence on personal relationships can sometimes lead to problems, however. A collaborative effort between industry and a university can founder if a key individual (the "champion" within the firm) is transferred or moved elsewhere within the company. Overall, therefore, the diversity of approaches to industry-academic collaboration is healthy for both partners.

Finding 2-4. Creating standards is an important aspect of innovation in communication systems.

Standards are necessary for interoperability, which is essential to the industry. The more interoperability, the faster the growth of the user base and the faster the increase in value of the system. The success of many businesses depends on the number of other entities that can communicate in a network—the value of network externalities.

Committees of industry members, sometimes with academic participation, usually determine standards. Many Internet standards groups, such as the Internet Engineering Task Force (IETF); the ATM Forum, which was organized to promote data-networking uses of Asynchronous Transfer Mode (ATM); and the discrete multitone modulation standard for asymmetric digital subscriber lines, have academic participants. In the United States, university researchers often have difficulty participating in setting standards because of time and travel demands. In Europe, academic participation has been stronger. U.S. researchers could be helpful in gathering data and analyzing standards proposals; good data and independent analysis could reduce squabbling over business biases and focus more attention on design issues.

Finding 2-5. Academic institutions have been at the forefront of network infrastructure deployment.

The United States led the way in deploying advanced infrastructure critical to enabling research (e.g., ARPAnet, NSFnet, vBNS, NGI, Internet2). The deployed infrastructure has led to further developments. For example, the World Wide Web was successful partly because networking infrastructure had already been installed.

Academic institutions have played a crucial role in the deployment of network infrastructure. Ever since NSFnet was formed, universities have recognized the importance of the Internet to academic endeavors of all kinds, not just computer science and engineering research. With funding from NSF and other sources, universities have been willing to deploy leading-edge technologies. Deployments such as Internet2, which was spurred in large part by universities, are likely to increase the speed and throughput of network services available to universities and thus to support research that requires high-performance networking infrastructure. However, this may not necessarily stimulate research on networking. Conflicting demands on these systems has created some tension between providing robust services for other research fields and experimentation for networking research. A state-of-the-art network that can be used for experiments in networking has not been developed.

Finding 2-6. The network systems and communications industry is evolving in directions that may require new kinds of university-industry partnerships to exploit research.

As communicating appliances proliferate, the need for harmonizing the technological and human elements increases. Examples include: designing communication services that users can understand and exploit; integrating multiple devices and services to create personalized configurations; designing new user interfaces; and streamlining or automating customer service functions. Even within a network, areas that are not purely technical will also require research: the provision of services; the quality of service; incompatibilities between heterogeneous products and services; security; and network management and operation. Optimal interactions among product design, network organization and management, service provision, and technology will require close collaboration between university researchers in the social, behavioral, and management sciences on one hand and engineers and scientists on the other. As the industry moves toward offering more "information services" rather than "communications devices," it must turn to the market rather than to research for guidance. In the future, university research might focus on how individuals and society as a whole value and use network systems and communications services.

Finding 2-7. Many Internet service providers are not willing to make their data available to researchers.

Many pressing research questions in network systems and communications require studies of the characteristics of networks under normal operating conditions as a basis of comparison. Research on limiting congestion, improving the quality of service, and improving routing requires traces or other logs of actual network activity. Although data are available for some experimental networks, many Internet service providers have not been willing to make their data available to researchers. This has hampered university research that might lead to improved network design and operation. The problem could get worse if service providers become more vertically organized and less open about their operations, problems, and needs.

Finding 2-8. The high cost of protecting intellectual property could impede research.

Industry and academia are both becoming increasingly protective of their intellectual property rights, and the enormous economic activity in the network systems and communications sector encourages this trend. The processes of working out licensing and sharing agreements could impede the free flow of ideas necessary for research to flourish. Whenever universities band together in a research consortium or a single university-industry collaboration is started, researchers spend much time and energy working out intellectual property agreements.

The trend can be counteracted in several ways. "Open source" software licenses that implicitly recognize that unused intellectual property has no value explicitly promote sharing. Other ways of reducing the costs of intellectual property protection could include standardized forms of collaborative research agreements.

Finding 2-9. Long-term research is important to the future of the network systems and communications industry.

Despite the apparent success of network systems and communications technologies, many difficult problems must be solved for the industry to continue to grow and prosper. Continued rapid expansion of a sophisticated communications infrastructure with millions or billions of network elements is bound to face some difficult problems. The industry needs software-engineering discipline to ensure that modules intended to fit together do so and that upgraded modules can be introduced without disrupting network operation. Distributed systems must be designed to be robust under failure, to remain stable under all operating conditions, and to guarantee performance requirements. Some long-term problems, such as the quality of service and security, are related to scale limitations. Current data networks require many people with sophisticated skills; and the design complexity and deployment scales of these systems exceed our engineering knowledge. As the extraordinary growth of the industry slows, new problems will have to be addressed, such as the impact of microelectronic components with different costs and technical properties, new computational needs as the rate of miniaturization slows, and the technical implications of these changes. Indeed, as industrial research investments change, and as human capital stresses wax and wane, it is important to keep long-term academic research activities alive, precisely because they are the long-lived seeds from which both ideas and people can spring, regardless of the short-term financial health of the industry.

Recommendations

Recommendation 2-1. Universities and industry should take steps to ensure that faculty and students are available to carry on research in computer science and other information technology fields in the future.

Innovation, either from research or incremental engineering, depends on trained researchers. Projected demand for computer science and other information technology graduates indicates periodic shortages in coming years. To maintain the pipeline of both academic and industrial researchers, the following measures could be taken:

- Universities should provide early research experiences for undergraduates or even secondary school students.
- Universities should provide career-development support for young faculty members.
- Fellowships should be provided for graduate students to encourage them to pursue research degrees; industry should provide some of this support.
- Universities and industry should provide incentives for industry engineers to return to academia for training in research.
- Universities should develop cooperative programs in which master's degrees are based not only on course work, but also on research experience.
- Training in academic research should include training in some of the qualities students will need for jobs in industry.
- Research should involve addressing not only small technical puzzles in isolation, but also complex systems problems in context. Students should be encouraged to confront complexity and to address real-world data and operational problems.
- Research should encourage teamwork.
- High-caliber industry researchers and engineers should be encouraged to take sabbaticals to work in academia, thus bringing real-world research problems into academic settings.

Recommendation 2-2. Universities and industry should continue to develop diverse collaborative arrangements.

Industry and universities should resist the temptation to impose standard structural mechanisms to promote collaboration. Incentives for personal interactions between university and industry should be encouraged in the following ways:

- Provide support for strong, committed leaders and the collaborative organizations they lead.
- Encourage sabbaticals in both directions, enabling academics to spend time in industry, especially in start-up companies.
- Support people and projects that involve academic and industry researchers in essential ways.
- Explore new ways to support personal interactions across academic-industry boundaries, including using technology to support collaboration.

Recommendation 2-3. Universities and industry should make every effort to invigorate academic research on networking.

The extraordinary success of the Internet and the lure of Internet-related start-up companies have tended to focus attention on short-term goals, causing long-term research to suffer. The situation could be improved in the following ways:

- Acknowledge that the research community must take risks.
- Focus academic research on the thorny problems of large systems: modeling, maintenance, upgrades, quality-of-service, security, and so on. Both funding agencies and academics must recognize that large-scale systems can best be addressed in a university setting. Even applied systems research can be structured in a way that accommodates a long-term approach.
- Universities and funding agencies (and industry) should support long-term, radical research on networks.
- Universities and industry should encourage interdisciplinary research that combines network technologies with design and social science disciplines. Networked devices (especially hand-held mobile devices) will have to meet both technical and human requirements.
- Universities should recognize that valuable innovations and engineering in the field are often not channeled through traditional peer-reviewed publications. Therefore, effective industry interaction should be more highly valued in decisions about academic promotion and tenure.
- To revitalize academic research on networking, the National Science Foundation should consider sponsoring a workshop on the subject that brings together academic and industry participants. A new agenda could provide a strong argument for industry support, either by individual firms or by a consortium.

NOTES

[1]Gordon Moore (cofounder of Intel) predicted in 1965 that the transistor density of semiconductor chips would double roughly every year. See Moore, 1965.

[2]For example, during fiscal year 2001, Microsoft spent $4.38 billion on product research and development activities excluding funding of joint venture activity. This represented 17.3 percent of revenue that year. Microsoft Research, the part of the company that looks more than one or two product cycles out, has around 600 employees and a budget of roughly $200 million, less than 5 percent of the $4.38 billion, or less than 0.8 percent of total revenue.

[3]The loss of faculty to commercial endeavors was limited in time and to only a few programs. Data from the most recent Taulbee Survey of computer science and computer engineering departments indicate that faculty numbers have grown and are anticipated to grow through 2004. The survey also indicates that faculty departures have ranged from 2.3 to 2.6 percent over the last several years (Bryant and Vardi, 2002).

[4]Economists have long acknowledged "externalities," factors that alter the value of a good viewed in isolation. Shapiro and Varian (1998) applied the idea to networks, so-called "network effects." Robert Metcalfe, a popular speaker on the value of networks, has often said that the usefulness, or utility, of a network equals the square of the number of users. This observation has been dubbed "Metcalfe's law" (Gilder, 1993).

REFERENCES

AT&T and the International Computer Science Institute. 1998. AT&T Labs, ICSI establish Internet research center. Press release October 8, 1998. Available online at: *http://www.icir.org/aciri.html.* [June 24, 2003]

Bailey, D.E., F.S. Settles, and D. Sanrow. 1998. Designing, Controlling, and Improving SRC Research Quality. Presentation at the NAE-Committee on Science, Engineering, and Public Policy Workshop on the Role of Human Capital in Capitalizing on Research, Beckman Center, Irvine, California, January 21, 1998.

BankBoston. 1997. MIT: The Impact of Innovation. Special Report of the BankBoston Economics Department. Available online at: *http://web.mit.edu/newsoffice/founders.* [June 24, 2003]

Bell, D.G., D.G. Bobrow, O. Raiman, and M.H. Shirley. 1997. Dynamic Documents and Situated Processes: Building on Local Knowledge in Field Service. Pp. 261–276 in Information and Process Integration in Enterprises: Rethinking Documents, T. Wakayama, S. Kannapan, C.M. Khoong, S. Navathe, and J. Yates, eds. Norwell, Mass.: Kluwer Academic Publishers.

Bryant, R.E., and M.Y. Vardi. 2002. 2000–2001 Taulbee Survey: Hope for More Balance in Supply and Demand. Computing Research News 14(2): 4–11.

Brynjolfson, E. 1991. Information Technology and the "Productivity Paradox": What We Know and What We Don't Know. Cambridge, Mass.: MIT Sloan School of Management.

Bureau of Labor Statistics. 2001. Table 3b. Fastest Growing Occupations, 2000–10. Available online at: *http://www.bls.gov/news.release/ecopro.t06.htm.* [June 24, 2003]

Card, S.K., T.P. Moran, and A. Newell. 1983. The Psychology of Human-Computer Interaction. Hillsdale, N.J.: L. Erlbaum Assoc.

Cerf, V., and R. Kahn. 1974. A protocol for packet network intercommunication. IEEE Transactions on Communications 22(5): 637–642.

Cohen, W.M., and D.A. Levinthal. 1990. Absorptive capacity: a new perspective on learning and innovation. Administrative Science Quarterly 35(1): 128–152.

CTIA (Cellular Telecommunications and Internet Association). 2003. CTIA's Semi-Annual Wireless Industry Survey. Available online at: *http://www.wow-com.com/pdf/CTIA_Survey_Yearend_2002.pdf.* [June 24, 2003]

CTIS (Corporate Technology Information Services, Inc.). 1999. Spotlight. Woburn, Mass.: Corporate Technology Information Services, Inc.

Cusumano, M. 1991. Japan's Software Factories: A Challenge to U.S. Management. Oxford, U.K.: Oxford University Press.

de Sola Pool, I., ed. 1977. The Social Impact of the Telephone. Cambridge, Mass.: MIT Press.

Earl, M., ed. 1988. Information Management: The Strategic Dimension. Oxford, U.K.: Clarendon Press.

Gilder, G. 1993. Metcalfe's Law and Legacy. Forbes ASAP, September 13. Available online at: *http://www.seas.upenn.edu/~gaj1/metgg.html.* [June 24, 2003]

Hill, S. 2001. Science and Engineering Degrees: 1966–1998. Arlington, Va.: Division of Science Resources Studies, National Science Foundation.

Intel. 2003. Intel Research Network of Laboratories. Available online at: *http://www.intel-research.net.* [June 24, 2003]

InternetWeek. 2001. Report: DSL Market to Rebound Next Year. October 24, 2001. Available online at: *http://www.internetweek.com/story/INW20011024S0005.* [June 24, 2003]

Kahn, A. 1970. The Economics of Regulation, Vol. 1. New York: John Wiley & Sons.

Kahn, A. 1971. The Economics of Regulation, Vol. 2. New York: John Wiley & Sons.

Keen, P.G.W., and M.S. Scott Morton. 1978. Decision Support Systems: An Organizational Perspective. Reading, Mass.: Addison-Wesley.

Kraut, R., W. Scherlis, T. Mukhopadhyay, J. Manning, and S. Kiesler. 1996. The HomeNet field trial of residential Internet services. Communications of the ACM 39(12): 55–63.

Lazowska, E. 1998. Remarks made during panel session on Changing the Interaction Between Academic Research and Industry: University, Industry, and Government Perspectives. Presented at the workshop How Can Academic Research Best Contribute to Network Systems and Communications?, National Academy of Engineering, Washington, D.C., October 30, 1998.

Malone, T. 1998. Remarks made during panel session on Contributions and Impact of Academic Research: Social, Management, and Policy Sciences. Presented at the workshop How Can Academic Research Best Contribute to Network Systems and Communications?, National Academy of Engineering, Washington, D.C., October 30, 1998.

Malone, T.W., K. Crowston, J. Lee, B. Pentland, C. Dellarocas, G. Wyner, J. Quimby, C.S. Osborn, A. Bernstein, G. Herman, M. Klein, and E. O'Donnell. 1999. Tools for inventing organizations: toward a handbook of organizational processes. Management Science 45(3): 425–443.

Markus, M.L. 1987. Toward a "critical mass" theory of interactive media: universal access, interdependence, and diffusion. Communications Research 14: 491–511.

McKnight, L.W., and J.P. Bailey, eds. 1997. Internet Economics. Cambridge, Mass.: MIT Press.

Messerschmitt, D.G. 1996. The convergence of telecommunications and computing: what are the implications today? IEEE Proceedings (August). Available online at: *http://www.informatik.tu-darmstadt.de/VS/Lehre/WS97-98/JavaTK/convergence.html.* [June 24, 2003]

MICRO. 2002. University of California Microelectronics Innovation and Computer Research Opportunities. Available online at: *http://www.ucop.edu/research/micro/.* [June 24, 2003]

Moore, G.E. 1965. Cramming more components onto integrated circuits. Electronics 38(8): 114–117. Available online at: *http://www.cybraryn.com/news/documents/moore-paper.pdf.* [June 24, 2003]

Morgan, R.P., and D.E. Strickland. 2000. U.S. university research contributions to industry: findings and conjectures. Science and Public Policy 28(2): 113–121.

Morris, J. 1998. Remarks made during panel session on Changing the Interaction Between Academic Research and Industry: University, Industry and Government Perspectives. Presented at the workshop How Can Academic Research Best Contribute to Network Systems and Communications?, National Academy of Engineering, Washington, D.C., October 30, 1998

NRC (National Research Council). 1995. Evolving the High-Performance Computing and Communications Initiative to Support the Nation's Information Infrastructure. Washington, D.C.: National Academy Press.

NRC. 1996. Cryptography's Role in Securing the Information Society. Washington, D.C.: National Academy Press.

NRC. 1999. Funding a Revolution: Government Support for Computing Research. Washington, D.C.: National Academy Press.

NRC. 2000. Making IT Better: Expanding Information Technology Research to Meet Society's Needs. Washington, D.C.: National Academy Press.

NRC. 2003. Innovation and Information Technology. Washington, D.C.: National Academies Press.

NSF (National Science Foundation). 2001. Survey of Federal Funds for Research and Development: Fiscal Years 1999, 2000, and 2001. Arlington, Va.: National Science Foundation.

Orr, J. 1990. Sharing Knowledge, Celebrating Identity: War Stories and Community Memory in a Service Culture. Pp. 169–189 in Collective Remembering: Memory in Society, D.S. Middleton and D. Edwards, eds. Beverly Hills, Calif.: Sage Publications.

Parker, L. 1997. The Engineering Research Centers Program: An Assessment of Benefits and Outcomes. Arlington, Va.: National Science Foundation.

Qualcomm Corporation. 1999. ERICSSON and QUALCOMM Reach Global CDMA Resolution. Press release. March 25, 1999. Available online at: *http://www.qualcomm.com/press/pr/releases1999/press457.html*. [June 24, 2003]

Rockart, J.F. 1981. The changing role of the information systems executive: a critical success factors perspective. Sloan Management Review 22(2): 15–25.

Roessner, D., R. Carr, I. Feller, M. McGeary, and N. Newman. 1998. The Role of NSF's Support of Engineering in Enabling Technological Innovation: Final Report-Phase II. Arlington, Va.: SRI International. Available online at: *http://www.sri.com/policy/stp/techin2/*. [June 24, 2003]

Schofield, J.W. 1995. Computers and Classroom Culture. New York: Cambridge University Press.

Shapiro, C., and H. Varian. 1998. Information Rules: A Strategic Guide to the Network Economy. Cambridge, Mass.: Harvard Business School Press.

Siegel, J.L., V. Dubrovsky, S. Kiesler, and T. McGuire. 1986. Group Processes in Computer-Mediated Communication. Organizational Behavior and Human Decision Processes 37: 157–187.

Sirbu, M. 1998. Remarks made during panel session on Contributions and Impact of Academic Research: Social, Management, and Policy Sciences. Presented at the workshop How Can Academic Research Best Contribute to Network Systems and Communications?, National Academy of Engineering, Washington, D.C., October 30, 1998.

Sproull, L., and S. Kiesler. 1991. Connections: New Ways of Working in the Networked Organization. Cambridge, Mass.: MIT Press.

SRI International. 1997. The Role of NSF's Support of Engineering in Enabling Technological Innovation. IV. The Internet. Available online at: *http://www.sri.com/policy/stp/techin/inter1.html*. [June 24, 2003]

Stanford University Corporate Guide. 2001. Top 10 Stanford Inventions. Available online at: *http://corporate.stanford.edu/innovations/invent.html*. [June 24, 2003]

Tennenhouse, D. 1998. Diagram presented during panel session on Changing the Interaction Between Academic Research and Industry: University, Industry and Government Perspectives at the workshop How Can Academic Research Best Contribute to Network Systems and Communications?, National Academy of Engineering, Washington, D.C., October 30, 1998.

Turkle, S. 1984. The Second Self: Computers and the Human Spirit. New York: Simon & Schuster.

U.S. Bureau of the Census. 2002. Statistical Abstract of the United States–2002. Washington, D.C.: U.S. Government Printing Office. Available online at: *http://www.census.gov/prod/www/statistical-abstract-02.html*. [June 24, 2003]

ADDENDUM

E-Mail Questionnaire

The following questionnaire was sent to individuals selected from various parts of the network systems and communication industry, some of whom attended the October 1998 workshop. Included among the questionnaire respondents were senior executives at AT&T Laboratories, Bell Atlantic, Bellcore, MCI, and Motorola, and professors with expertise in computer science and engineering, network systems, and telecommunications from Stanford University, University of Delaware, University of California-Berkeley, University of California, Los Angeles, University of Virginia, and University of Washington.

THE IMPACT OF ACADEMIC RESEARCH ON INDUSTRIAL PERFORMANCE
NETWORK SYSTEMS AND COMMUNICATIONS PANEL

We invite your responses to the following questions. Your responses will be used by our Panel as background information for our report. Any material used verbatim will not be attributed to you without seeking your permission.

1. Could you describe briefly significant academic research contributions to the network systems and communications industry? (If possible, please supply references to published information that outlines the contributions.)

2. Overall, would you describe the impact of academic research on industrial performance in the network systems and communications industry as (Please put an X in one box):

☐ 1. very large
☐ 2. large
☐ 3. medium
☐ 4. small
☐ 5. very small/non-existent

3. What is the role of academic *research* in educating people who work in your industry? (Please focus on university *research* activities, rather than university education generally.)

4. What structural forms of university-industry collaboration lead to good results in your industry? An example of such a structure might be a discipline- or industry-oriented "center" that solicits industry sponsors for a collection of

projects that span a varied research program. What seem to be the essential determinants of success of such structures?

5. What are significant emerging trends or problems that the network systems and communications industry will face in the future that could benefit from academic research?

6. What changes are required, if any, in academic research if it is to be responsive to these industrial trends and problems?

7. What single step could be taken by universities to enhance the impact of academic research on the industry?

8. What single step could be taken by companies to enhance the impact of academic research on industry?

9. What single step could be taken by government to enhance the impact of academic research on industry?

10. Do you see any downside to enhanced university-industry research collaboration? Things to be avoided?

11. Other comments? Any comments, pointers to other studies, or suggestions would be appreciated.

WORKSHOP AGENDA

HOW CAN ACADEMIC RESEARCH BEST CONTRIBUTE TO NETWORK SYSTEMS AND COMMUNICATIONS?

October 30, 1998

National Academies Building
2101 Constitution Avenue N.W.
Washington, D.C.

9:00 am **Welcoming remarks and self-introductions**
Wm. A. Wulf, President, National Academy of Engineering

9:15 am **Overview of the work of the Network Systems and Communications Panel and** description of the wider NAE study
Bob Sproull, Panel Chair

10:00 am Break

10:15 am **Session I. Contributions and impacts of academic research on performance in the network systems and communications industry: Engineering and the Physical Sciences**
David Forney, Ambuj Goyal, Robert Kahn, H.T. Kung, David Mills

11:45 am Lunch in Meeting Room

12:30 pm **Session II. Contributions and impacts of academic research on performance in the network systems and communications industry: Design, Social, Management, and Policy Sciences**
Dan Atkins, Walter Bender, Robert Kraut, Tom Malone, Marvin Sirbu

1:30 pm **Session III. Structures for university-industry collaboration**
James Flanagan, Stewart Personick, David Roessner, Donald Strickland, Stephen Wolff

2:30 pm Break

2:45 pm **Session IV. Changing the interaction between academic research and industry: University, Industry, and Government Perspectives**
Hamid Ahmadi, Ed Lazowska, James Morris, Rick Rashid, George Strawn, David Tennenhouse

4:30 pm **Discussion, conclusions and recommendations**
Bob Sproull

WORKSHOP ATTENDEES

Robert Sproull, *chair* *
Vice President and Sun Fellow
Sun Microsystems, Inc.

Hamid Ahmadi
AT&T Labs

Alfred V. Aho *
Associate Research Vice President
Communications Sciences
Research Division
Lucent Technologies
Bell Labs Innovations

Daniel Atkins
School of Information
University of Michigan

Walter Bender
MIT Media Lab

John Cioffi *
Associate Professor
Department of Electrical Engineering
Stanford University

David J. Farber *
Alfred Fitler Moore Professor of
Telecommunications
University of Pennsylvania

James Flanagan
Center for Computer Aids
Rutgers University

G. David Forney, Jr.
Motorola, Inc.

Ambuj Goyal
IBM Corporation
T.J. Watson Research Center

George H. Heilmeier *
Chairman Emeritus
Bellcore

Robert Kahn
Corporation for National Research
Initiatives

Robert Kraut
Department of Social and Decision
Sciences
Carnegie Mellon University

H.T. Kung
Department of Electrical Engineering
and Computer Science
Harvard University

Ed Lazowska
Department of Computer Science and
Engineering
University of Washington

Tom Malone
Sloan School of Management
Massachusetts Institute of Technology

David Mills
Department of Electrical and
Computer Engineering
University of Delaware

*Panel member

James Morris
School of Computer Science
Carnegie Mellon University

Stewart Personick
Drexel University

Richard Rashid
Advanced Technical Research
Microsoft

J. David Roessner
School of Public Policy
Georgia Institute of Technology

Jerrard Sheehan
National Research Council

Marvin Sirbu
Information Networking Institute
Carnegie Mellon University

George Strawn
National Science Foundation

Donald E. Strickland
Chair, Management Department
Southern Illinois University

David Tennenhouse
Defense Advanced Research Projects Agency

Stephen Wolff *
Executive Director
Advanced Internet Initiatives Division
Cisco Systems, Inc.

Wm. A. Wulf
President
National Academy of Engineering

NAE Program Office Staff

Tom Weimer, Director
Proctor Reid, Associate Director
Nathan Kahl, Project Assistant
Robert Morgan, NAE Fellow and Senior Analyst

*Panel member

3

Report of the Panel on the Medical Devices and Equipment Industry

Rapid changes in the financing and delivery of U.S. health care may have a significant effect on the incentives for universities and industrial firms to generate, evaluate, and introduce new medical devices. This report examines the interactions of these two critical participants in technological changes, specifically the contributions of academic research to the medical device industry. The Panel on Medical Devices and Equipment, one of the five panels formed by the Committee on the Impact of Academic Research on Industrial Performance of the NAE, hopes this report will provide a starting point for further research on critical, but often neglected, institutional interactions in the medical device innovation process.

The Panel on Medical Devices and Equipment comprised six members: one National Academy of Sciences member from academia, one Institute of Medicine member from industry, one other member from academia, and three more from industry. Three of the panel members were also members of the parent committee. The panel assessed the contributions of academic research, which may include new knowledge, inventions, and the training of people in modern research techniques, to the medical device industry and recommended ways of improving such contributions in the future. This assessment is especially timely in view of the fundamental changes occurring in the American health-care system, including academic medicine, and American higher education, which are putting unprecedented pressures on both academic medical centers and medical device firms. In the course of this study, the panel reviewed the literature, developed several case studies, and sent a questionnaire to individuals in academia, the medical device industry, and government. This questionnaire was followed by a workshop attended by approximately 35 senior individuals in the medical device sector (see Addendum).

There are several compelling reasons for undertaking a close examination of the interface between firms and universities in the medical device sector. First, although this industry, like the pharmaceutical industry, develops and markets products that contribute to human health and well-being, it has received far less attention than the pharmaceutical industry. Second, the number and variety of interactions between universities and industrial firms has increased significantly. Third, a common misperception of the relationship between industry and universities assigns to universities the role of generating fundamental (basic) knowledge and to industry the role of performing applied research and developing medical technologies. A closer look at the ways medical innovations arise and spread suggests that both parties perform much more complex, subtle, and wide-ranging roles than conventional wisdom suggests.

This report addresses two sets of questions:

- What role has university-based research played in technological advances in the medical device industry? What impact has academic research had on the industry's performance? Are the current mechanisms for university-industry collaboration, both formal and informal, adequate?
- How might academic research contribute more effectively to the medical device industry in the future? Are there new modes of university-industry collaboration that would increase the payoffs without compromising the core mission of either sector? What specific actions might increase the contributions of academic research to the industry's performance?

Whereas the focus of the report is on the contributions of academic research to industry, important contributions also flow in the other direction. Industry, among others, contributes resources for conducting university R&D and for training students. Both academic and industrial institutions are involved in the whole innovation cycle—research, development, manufacturing, evaluation, marketing, and product modification. Industry and universities have distinctive, complementary skills, as well as overlapping competencies. In fact, one characteristic of innovations in medical devices is close collaboration, even codependency, between universities and industry firms.

The first part of this report is a review of the main components and a definition of the boundaries of the medical device industry. Following a brief overview of the structural and performance characteristics of the industry, the main players in the innovation system for medical devices are identified, and the multifaceted nature of research relations between academia and the medical device industry are analyzed. Sweeping changes are occurring in the health-care environment, including the introduction of market forces and the widespread diffusion of managed care into the delivery of health care, modifications in Food and Drug

Administration (FDA) regulations, and new policies and practices regarding intellectual property rights. This report attempts to weigh the effects of these changes on university-industry relations and consider how university contributions to the medical device industry in this rapidly changing environment could be enhanced.

DEFINITION OF THE INDUSTRY

Main Components

Medical devices encompass a heterogeneous group of products, ranging from low-tech, inexpensive devices, such as tongue depressors and disposable needles, to sophisticated devices, such as implanted therapeutic devices, lithotripters, and magnetic resonance imaging (MRI) machines. The U.S. Department of Commerce currently groups medical devices into five categories, according to North American Industrial Classification System (NAICS) codes:

- **Surgical and medical instruments** (NAICS 339112) include medical, surgical, ophthalmic, and veterinary instruments and apparatuses. Examples are syringes, hypodermic needles, anesthesia apparatuses, blood transfusion equipment, catheters, surgical clamps, and medical thermometers.
- **Surgical appliances and supplies** (NAICS 339113) include orthopedic devices, prosthetic appliances, surgical dressings, crutches, surgical sutures, and personal industrial safety devices (except protective eyewear).
- **Dental equipment and supplies** (NAICS 339114) include equipment and supplies used by dental laboratories and dentists offices, such as chairs, instrument delivery systems, hand instruments, and impression materials.
- **Irradiation apparatuses** (NAICS 334517) include apparatuses used for medical diagnostic, medical therapeutic, industrial, research, and scientific evaluations.
- **Navigational, measuring, electromedical, and control instruments** (NAICS 334510) include electromedical and electrotherapeutic apparatus, such as MRI equipment, medical ultrasound equipment, pacemakers, hearing aids, electrocardiographs, and electromedical endoscopic equipment.

This report focuses mainly on the high-tech, innovation-driven segments of the industry, such as implantable devices, bioengineered devices, optical instruments, surgical staplers, and surgical miniaturization, in which the contributions of academia are likely to be most apparent. Most FDA Class 3 devices, for which sponsors are required to demonstrate safety and efficacy before the FDA grants marketing clearance, are included in this category. It also includes so-called "510(k) devices," which are "substantially equivalent" to devices that were on the market before the 1976 Medical Device Amendments took effect and, therefore, are subject to less stringent regulatory review. This study also examines emerging market segments that are expanding the boundaries of the traditional device industry, such as tissue engineering and health information systems intended to improve the quality and efficiency of health-care delivery systems.

The global market for medical devices was $138 billion in 1999. The U.S. market accounts for 37 percent of global demand, and the U.S. industry supplied 40 percent of the global market with shipments of $55 billion in 1999 (McGraw-Hill and U.S. Department of Commerce, 2000). The United States traditionally runs a positive balance of trade in medical device products (estimated to be $7 billion in 2000), and several American firms have strong market shares in Europe and Japan (AdvaMed, 2001). All major firms throughout the world participate in the U.S. market; most leading foreign firms have U.S. sales subsidiaries, and many also have extensive research and manufacturing activities in the United States. As of 1999, both domestic and foreign medical device firms operating in the United States employed almost 300,000 workers, and the medical device industry was one of the fastest growing manufacturing sectors in the U.S. economy (U.S. Bureau of the Census, 2001).

Companies in this industry capture relatively few sales from any single product. The norm for important therapeutic tools (e.g., vascular grafts) is a total global market of $70 million (Wilkerson Group, 1995). Even "blockbuster" products rarely surpass $100 million. The Johnson & Johnson Palmaz-Schatz stent for coronary heart disease was unusually successful in garnering sales of almost $400 million annually in its early years. But despite the purported strength of the Johnson & Johnson patent and its headstart in the market, new companies continued to improve stent designs for opening coronary arteries. As a result, Johnson & Johnson lost more than 70 percent of its market share in five years to new entrants. Johnson & Johnson is expected to make a comeback, however, because of sharply reduced restenosis with its new drug-coated stents.

In short, this is a dynamic industry driven by intense competition. Products that briefly capture sales are swept away within a few years by more innovative replacements. Consequently, research activity is intense; publicly traded device firms invest 12.9 percent of sales in R&D, and the most innovative firms reinvest as much as 23 percent of sales revenues in R&D. This figure is comparable to investments by aggressive pharmaceutical companies (Lewin Group, 2000).

The Roles of Large and Small Firms

The extremely diverse medical device industry includes small start-up companies and giant corporations. In 1999, 65 percent of firms had fewer than 20 employees, and only 12 percent had more than 100 (U.S. Bureau of the Census, 2001). The correlation between the size of a firm and its role in the market is not entirely clear, but it is widely believed that small firms play a disproportionate role in initial innovation and that large firms determine the commercial success of new products.

Large Firms

The underlying economics of the industry drive product development toward larger firms that have the sophisticated assets to exploit the commercial potential of innovations and can navigate the complex regulatory requirements for the introduction of new health-care products. First, as a result of multiple filings, large firms have developed the capability of managing clinical trials to meet regulatory requirements. An excellent example is the development of the home HIV test. Although the technology was relatively simple, numerous start-up companies had failed to demonstrate their ability to collect and test potentially contaminated blood in the home setting. Johnson & Johnson, which has extensive knowledge of the regulatory process, was able to shepherd the first successful home HIV diagnostic test to market. Second, large companies often have considerable skills in manufacturing and marketing. The history of diagnostic imaging, for example, clearly shows that first-mover advantages are not always a key to success in the marketplace of new technologies that have significant commonalities with earlier technologies (e.g., MRI with CT). Although large multinational companies were often late entrants, their skills in marketing and servicing and their established reputations often enabled them to assume dominant positions (Gelijns et al., 1998). Third, large, experienced companies understand the purchasing patterns of multiple stakeholders in a complex hospital environment. Because buyers prefer to contract with a limited number of suppliers, successful device companies offer a full product line of compatible products. Small companies with the most innovative devices may gain a foothold but can rarely maintain it.

Finally, the most successful companies plan for short product life cycles, and they swiftly introduce incremental enhancements developed by internal R&D. These companies rarely invest in basic research because the direct returns on basic science are relatively low during the short payback time for internalizing and commercializing product concepts. Consequently, larger firms invest in sophisticated market scanning and acquisition capabilities to identify new ideas for internal development and tend to leave "breakthrough innovations" to others. To be sure, large companies do produce pioneering innovations from internal research, but these breakthroughs often leverage technologies from preexisting products. In addition, large companies can exploit the experiences of users to produce next-generation products.

Innovative Small Firms

The existence of numerous small, innovative start-up companies in the medical device industry has been well documented. A study of publicly traded medical device firms found that in 1997, 65 percent of firms had fewer than 50 employees. Firms with less than $5 million in revenue spent an average of 252 percent of sales revenue on R&D (Lewin Group, 2000). These research-focused companies specialize in the "front end" of R&D, and perhaps not surprisingly, a study by the Wilkerson Group concluded, "nearly all significant new and innovative products and procedures were pioneered by start-up companies." Indeed, in their survey they cite 29 major therapeutic advances, all of which could be attributed to start-ups (Wilkerson Group, 1995).

According to Gelijns and colleagues (1994):

> Attempts to measure the innovative activity of [medical device] firms as a function of their size have long been handicapped, not only by methodological but by conceptual difficulties—for example, the absence of an unambiguous criterion for recognizing and, therefore, for measuring innovations, or for distinguishing between "major" and "minor" innovations. One study conducted in the early 1980s by the Futures Group defined large firms as having more than 500 employees and small firms as having fewer than 500 employees (OTA, 1984). The same study reviewed more than 8,000 innovations published in trade journals in 1982 (which were likely to have overstated the contributions of large firms and understated the contributions of small firms) and calculated rates of innovation per employee for each of the 5 SIC (now NAICS) code medical device categories. The study concluded that, with the exception of the small ophthalmic goods category, small firms were more than twice as innovative per employee as large firms (OTA, 1984). This conclusion reflects the likely differences in the workforces of small and large firms; small firms are often "R&D boutiques" that do not have large numbers of personnel in, for instance, regulatory affairs, marketing, or distribution infrastructure that large firms have.

The medical device industry also relies on individual inventors for ideas for new products. Once a working prototype or proof-of-concept device has been produced, the inventor is in a position to negotiate with large companies for a license or to create a start-up company. Besides individual initiative, small companies capable of demonstrating product potential require venture capitalists, high-risk/high-return investors willing to bankroll entrepreneurs with unproven technology.

INNOVATION SYSTEM

The medical device industry depends heavily on an infrastructure of institutions and activities outside the industry. Traditionally, both large and small firms have depended heavily on nonmedical industry sectors (e.g., those that deliver customized components or highly specialized materials), as well as research universities, especially academic medical centers (AMCs).

Government policies have also had a strong impact on innovation practices and university-industry relationships. First, although only a modest percentage of the federal budget is allocated directly for R&D on medical devices, the federal government is a major source of R&D funding. Second, the federal government influences the development process through the FDA's premarketing approvals and policies for medical devices. Third, the government has become a major source of payments to the providers of medical care (e.g., Medicare). For example, by including end-stage renal disease as coverable by Medicare, the government assured a market, which led to significant innovations in exchange devices, biocompatible materials, and other technologies necessary for dialysis. Government decisions have a decisive influence over how existing technologies are used. In addition, government decisions have a powerful impact on the financial incentives for private industry to undertake R&D.

Government is not the only payer that influences the market for medical devices. In recent years, managed care, ranging from classic health maintenance organizations (HMOs) to modified fee-for-service programs, has grown rapidly. Managed-care purchasers are taking a more critical and more independent stance about which technologies they will cover and the level at which they will reimburse providers; thus, they too influence the demand for new technologies.

Research Universities and Academic Medical Centers

Research universities are key players in the medical device innovation system. Basic advances in physics, materials sciences, optics, analytical methods, and computer science have resulted in many new device capabilities. Bioengineering research has emerged as a separate discipline in the last few decades; in 1998, 70 universities and colleges offered degrees in bioengineering.

A typical AMC generally comprises a medical school, a teaching hospital, a network of affiliated hospitals, and a nursing school. Some AMCs also have schools of dentistry, schools for allied health professionals, and schools of public health. These complex, multifunctional organizations have a three-pronged mission: (1) training clinicians and biomedical researchers, thereby ensuring the distribution of medical skills and specialties; (2) providing advanced specialty and tertiary care and therefore adopting the latest technologies; and (3) conducting biomedical research, ranging from laboratory-based fundamental research to population-based clinical studies.

In the United States, AMCs, and biomedical research in particular, have been major beneficiaries of post-World War II science policy. Total national investment in health-related R&D (public and private) has increased dramatically in the postwar period, more than three-fold since 1985 to more than $42 billion in 1998 (Commonwealth Fund Task Force on Academic Health Centers, 1999). At the same time, health insurance coverage was expanded, and Medicare was established. Medicare pays AMCs for patient care and educational activities.

These financial incentives encouraged the spectacular growth of American academic medicine. Between 1960 and 1992, the average medical school budget in the U.S. expanded nearly 10-fold in real terms (see Table 3-1). The table shows

TABLE 3-1 The Growth of U.S. Academic Medicine, 1960–1992 (in 1992 dollars)

	1960	1970	1980	1992
Support from NIH ($ millions)	1,320	3,028	5,419	8,407
Average medical school budget ($ millions)	24.1	64.6	91.9	200.4
Full-time medical school faculty (no.)				
Basic	4,023	8,283	12,816	15,579
Clinical	7,201	19,256	37,716	65,913
Matriculated medical students (no.)	30,288	40,487	65,189	66,142

SOURCE: Adapted from Iglehart, 1994.

that basic science faculty increased from 4,023 to 15,579, and clinical faculty increased far more rapidly from 7,201 to 65,913 over the three-decade period (Iglehart, 1994). As of the late 1990s, about 30 percent of all health-related R&D in the United States took place at AMCs (Commonwealth Fund Task Force on Academic Health Centers, 1999). Clinical specialists are major participants in the clinical testing and advancement of devices.

The financial support structure for AMCs, which is quite different from the support structure for other components of universities, contributed significantly to their past research success; AMCs have also developed a separate culture (Keller, 1998). AMC research activities are funded by a variety of sources. The federal government has funded the majority of AMC research (nearly 70 percent), especially for basic biomedical research. In 1996, the government funded more than 70 percent of new AMC research projects and more than 60 percent of all new nonclinical research or research on nonhuman subjects (Director's Panel on Clinical Research, 1997). In recent years, funding for academic research has increased under a variety of arrangements. Foundations and philanthropical organizations are also important sources of research funding. A substantial portion of academic research is funded internally; revenues from faculty practice plans, for example, are often used to underwrite research (they support about 9 percent of research, mostly clinical, in AMCs). An analysis in 1999 of six AMCs showed that, on average, clinical enterprises transferred about $50 million a year to medical schools for academic purposes. Universities also provided institutional funding to support the direct costs of research (Commonwealth Fund Task Force on Academic Health Centers, 1999).

Federal Agencies

Federal support for R&D in medical devices flows through multiple institutional and disciplinary channels. Although the majority of medical device-related

R&D funds is spent in AMCs, federal agencies also fund basic and applied research in academic science departments and engineering schools, federal laboratories, and industry proper.

The United States spends a larger percentage of its federal research budget on research in the life sciences than any European country (NSF, 2000). Between 1985 and 2001, federal obligations for research in the life sciences more than doubled, totaling more than $18 billion in 2001 (NSF, 2001). Most of NIH's overall budget of more than $13 billion a year is spent on extramural research at AMCs, particularly in basic (nonhuman subjects) research. Only a small portion of NIH's budget is allocated specifically to create opportunities for the development of devices. For example, in 1964 the National Heart, Lung, and Blood Institute (NHLBI) created the artificial heart program to support the development of a family of devices to assist patients with failing hearts and to rehabilitate patients with heart failure (Watson et al., 1994). NHLBI has also invested in clinical trials of cardiac devices, for example, to determine the effectiveness of defibrillators in high-risk patients with coronary artery disease and in the left ventricular-assist device for end-stage patients with heart failure.

Determining the portion of the NIH budget directly related to R&D on medical devices, however, is problematic. A congressional study in 1992 estimated that the government had invested about $422 million in R&D on medical devices (Littell, 1994). A 1998 report in *Science* estimated that NIH funding of bioengineering-related research projects, including biomaterials, prosthetic devices, and artificial organs, amounted to $417 million in 1996 (Agnew, 1998); this figure increased to $500 million in 1998 (Chronicle Information Resources, 1999). In 2000, NIH created the National Institute for Biomedical Imaging and Bioengineering (NIBIB) to "improve health by promoting fundamental discoveries, design and development, and translation and assessment of technological capabilities. The Institute will coordinate with biomedical imaging and bioengineering programs of other agencies and NIH institutes to support imaging and engineering research with potential medical applications and will facilitate the transfer of such technologies to medical applications" (P.L. 106-580). NIBIB's FY02 budget was $112 million.

In addition, the government supports some applied research in industry settings. In the early 1980s, for example, the federal government established the Small Business Innovation Research (SBIR) Program; and in 2000, 10 federal agencies awarded $1.1 billion in SBIR grants. Since the program's inception in 1983, the life sciences, which include medical devices, have received more than $2 billion in awards from NIH. NIH's SBIR awarded $435 million in 2001 (Goodnight, 2002).

Food and Drug Administration

The introduction of new or modified medical devices is subject to stringent and complex regulations. The Medical Device Amendments of 1976 were intended to ensure that new devices are both safe and effective before they are marketed. These amendments divide medical devices into three classes, depending on their potential risks to patients.

Approximately 30 percent of all medical devices are grouped in Class 1, which comprises instruments (e.g., stethoscopes) that do not support or sustain human life and do not present an unreasonable risk of illness or injury. Class 1 devices are subject to the general controls used before passage of the Medical Device Amendments—for example, regulations regarding registration, premarketing notification, record keeping, labeling, and good manufacturing practices. About 60 percent of devices fall into Class 2, which includes x-ray devices and other devices that pose some risk. Class 2 devices are subject to federally defined performance standards. Class 3 devices include all life-supporting or life-sustaining devices that substantially prevent health problems or that could pose a risk of injury or illness. For Class 3 devices, the sponsor must demonstrate safety and efficacy before the FDA grants marketing approval. Approximately 10 percent of medical devices fall into Class 3; examples include left-ventricular assist devices and laser angioplasty devices. Since 1976, all new devices are automatically placed in Class 3 unless the sponsor successfully petitions the FDA to reclassify them as "substantially equivalent" to a device that was on the market before the amendments took effect. Demonstration of this equivalence, called a 510(k) submission, is provided by descriptive, performance, and even clinical data.

To support a marketing approval decision, or in some instances a 510(k) submission, a sponsor must conduct clinical studies. If a device poses a significant risk, the sponsor submits a request for an investigational device exemption (IDE) to the FDA. Following clinical studies, the device may be approved for marketing through a so-called premarket approval (PMA) decision. Most PMAs are individual licenses secured by the developer for particular devices and specific clinical uses or indications. Other developers of similar kinds of devices must submit separate PMAs and clinical data.

In the 1990s, FDA regulation of medical devices changed significantly with the passage of the Safe Medical Devices Act of 1990. Under new requirements for premarketing studies, manufacturers are required to conduct more rigorous studies with appropriate, and where possible randomized, controls. Postmarketing surveillance now provides a number of separate mechanisms for collecting data. Both device manufacturers and health-care providers must report information indicating if the device may have caused or contributed to a death or serious injury. For high-risk devices, companies must keep track of patients, and, in certain cases, must conduct postapproval clinical studies to detect possible risks associated with the use of the device, as well as information on its effectiveness. These changes should improve the quality of evaluations and provide more information about safety and efficacy. At the same time, they have slowed the pace of introductions of new medical devices and increased the risk and cost for medical device firms.

In the early 1990s, the FDA had long review times for IDEs, PMAs, and 510(k) submissions, and the agency had accumulated a considerable backlog. Subsequently, the FDA reorganized its device branch, and then, in 1997, the FDA Modernization Act (also known as FDAMA) was passed. This wide-ranging legislation attempts to shift resources in the agency from relatively low-risk to relatively high-risk areas and to specify the requirements for trials of clinical devices. As a result of these changes, the backlog was diminished substantially and review time was shortened considerably; in 1998, for instance, the average review time for a 510(k) submission decreased by 12.3 percent from the preceding year.

Venture-Capital Industry

The United States has a mature venture-capital industry that provides access to liquid capital markets for the financing of high-risk ventures. Venture capital has been pivotal to the development of the industry, because the development and commercialization of medical devices can be a prolonged process, and few inventors can survive with debt financing alone. Venture capital allows the originator to obtain operating funds and to share financial risks.

Small firms with no track records often need multiple sources of funding for a substantial period of time, usually beginning with private financing and proceeding to the equity markets. Venture capitalists fund these companies when revenues are small or even nonexistent. As recently as 15 years, ago, the venture-capital market was small, but in 2001, health care, principally biopharmaceuticals and medical devices, received $5.6 billion in venture capital, 17 percent of total venture capital investments for the year (Zemel, 2002). In addition, the initial public offering (IPO) market expanded in the 1990s, which allowed venture capitalists to exit projects and thereby reap rewards for the risk they had borne; small companies subsequently had access to large pools of liquid capital for future expansion. IPO investment in the medical device industry rose from $410 million in 1995 to $1.268 billion in 1996; much of this growth was in the cardiovascular device sector. In recent years, with the economic downturn, venture-capital investment in medical devices has declined sharply, and IPOs have come to a virtual standstill. In 2001, there were only eight IPOs of medical device firms, raising roughly $741 million.

Third-Party Payers

In the last 20 years, dramatic changes have been made in the financing and delivery of U.S. health care. Changes include the rapid growth of managed care initiatives and the consolidation of hospitals and clinics into large integrated delivery systems. Managed care organizations increasingly reimburse health-care providers on a capitated basis (i.e., fixed reimbursement per patient per month), promote cost-conscious purchasing by negotiating price discounts on high-volume procedures, and may use selective contracting to concentrate sophisticated devices and related procedures in a smaller number of institutions. As a result, the incentives for industrial firms to generate new medical devices has also changed.

These changes have had many consequences for AMCs and university-industry interactions. First, the pressures to contain the costs of medical care have reduced the resources in AMCs for cross-subsidizing research. Second, researchers in AMCs in areas with high managed care penetration are less likely to obtain NIH grants or to publish than investigators in areas with less managed care competition (Griner and Blumenthal, 1998). The decrease is especially apparent in clinical research, raising questions about whether the necessary level of clinical research for the medical device enterprise can be sustained. Finally, the payment for treatment of patients in clinical trials is becoming increasingly contentious, which could inhibit the refinement of devices and the collection of data on medical devices.

Traditionally, industry has supported the evaluation of medical devices, government has supported the evaluation of major clinical procedures and off-label uses, and payers have supported (often unknowingly) the treatment costs of patients enrolled in clinical trials. However, managed care organizations have become increasingly reluctant to do so, and they are coming under increasing pressure to support these trials. Every sector—the federal government, industry, AMCs, and managed care—would obviously prefer that others shoulder more of the burden, but as both the number of evaluations and their complexity and sophistication have increased, the need for partnerships to pool resources has become evident. Conditional coverage, in which payers cover the costs of patient treatment in a predetermined research protocol while government and industry cover the costs of doing the research, is one option for intersector funding. In 1995, an Interagency Agreement between the FDA and the Healthcare Financing Administration made certain Category B nonexperimental/investigational devices eligible for Medicare payment during clinical trials. More than 90 percent of investigational device exemptions (IDEs) have been made eligible for Medicare payment in this manner. Legislation introducing conditional coverage for all Medicare enrollees was enacted in September 2000 (42 CFR411.1).

CONTRIBUTIONS OF ACADEMIC RESEARCH

Education and Training

One of the most important long-term contributions of academia is the training of people skilled in research techniques. Universities train people in many disciplines—biological, behavioral, and physical sciences, as well as engineering. Advances in the biological sciences, biomaterials sciences, and in biological information processing and analysis have ushered in a new era of progress and innovation in medical devices and bioengineering. In the 1990s, bioengineering was the fastest growing specialty at engineering schools that offer these programs (Agnew, 1998). To meet the growing need for multidisciplinary education in engineering, biology, and medicine, beginning in the late 1960s various educational programs have been developed to integrate engineering and clinical medicine. The University of Pennsylvania, Johns Hopkins University, Purdue University, and Rice University are some of the institutions that had early multidisciplinary programs. NIH and NSF have funded the development of more of these programs.

Medical schools train clinicians in a wide range of specialties, as well as scientists in laboratory-oriented basic research, translational (applied) research, and clinical evaluative sciences. In the 1970s and 1980s, NIH funding was increased for basic, laboratory-oriented research, which was reflected in an increase in the number of Ph.D.s receiving NIH awards and a decrease in the number of physician/scientists receiving awards. Since then, the ratio of Ph.D. applicants to physician applicants has remained constant at 3:1 (Commonwealth Fund Task Force on Academic Health Centers, 1999). Between 1994 and 1996, as pressures to contain costs increased and demands on physicians to maintain a certain volume of clinical care mounted, the number of first-time physician applicants to NIH dropped by 30 percent, raising concerns that the number of clinical researchers might be permanently diminished. As a result, NIH created new training (and research) initiatives for clinical researchers, in both translational and clinical evaluative research (e.g., biostatistics, clinical epidemiology, and outcomes research, fields that were traditionally covered by schools of public health). The number of training programs leading to joint MD/Masters of Public Health (MPH) degrees have increased as a result.

Schools of public health train people in sociomedical, health management, and policy sciences. As device production, evaluation, and marketing become more difficult, industry demand for people trained in outcomes analysis, biostatistics, health economics, and medical decision analysis has increased. Although educational programs in these areas have been established around the country, many of them are small, have insufficient clinical involvement, and are not geared toward the assessment and regulatory approval of medical devices. Business schools, which train people in the management sciences, have expanded programs that offer MD/MBA programs.

Research in the Physical Sciences and Engineering

Because markets for medical devices are often fragmented and relatively small, the medical device industry historically has not invested heavily in basic research but has depended heavily on scientific and technological capabilities developed in other sectors. Medical devices have exploited research and new technological capabilities and components developed by universities, the military, the electronics industry, and firms that manufacture specialized materials, such as high-quality glass for fiber optics and special materials for prosthetic devices.

Arguably, there have been two very distinct patterns to the collaborations between physical scientists and engineers on the one hand and clinical researchers on the other. One is in the field of instrumentation where electrical engineers and physicists either had separately developed a technology that could be used in a device or instrument or, working on a problem defined by a clinical researcher, had come up with a device or instrument that would solve the problem. Earl Bakken's development of the pacemaker is an example of the latter.

However, a very different kind of collaboration developed between mechanical and chemical engineers and clinical researchers, in which the engineers became directly involved in defining the problem, not merely helping to find the solution. This manifested itself in studies of fluid mechanics and transport phenomena in blood flow, and in the many studies aimed at characterizing the interactions between biological fluids and synthetic materials. These collaborations led to insights that have been key to understanding the effect of flow patterns in certain diseases, like atherosclerosis, but also to understanding the importance of flow patterns in prosthetic devices in promoting or inhibiting thrombosis or hemolysis. In these cases, engineers did not borrow from other fields but became involved in direct research in the biological systems to understand the unique phenomena of those systems.

In a 1998 study of trends in medical device technology conducted by the FDA, a survey revealed six somewhat overlapping "trend categories": computer-related technology; molecular medicine; home care and self-care; minimally invasive procedures; combination drug/device products; and organ replacements and assist devices (Herman et al., 1998). Most of these categories reflect contributions either from universities or other industry sectors. For example, computer-related technologies, which include computer-aided diagnosis, intelligent devices, biosensors and robotics, and networks of devices, all depend on the results of R&D in computers and communications, which, in turn, interact with and build on advances in mathematics, computer science, electrical engineering, and other disciplines. Minimally invasive procedures include minimally invasive instruments, medical imaging, microminiaturized devices, laser diagnosis and therapy, robotic surgical devices, and nonimplanted sensory aids, all of which make use of developments in physics, mathematics, electrical engineering, and computer science. Organ replacements and assist devices depend on advances in the materials sciences and, increasingly, on the interface between biology and the physical and engineering sciences.

Universities play an important role in the evolution of medical devices. University research may lead to the discovery of new scientific or technological principles, new designs, new materials, and advances in computer sciences. Examples of academic contributions and university-industry collaborations can be found in the development of lasers, endoscopy devices, and medical imaging machines. For instance, work by Charles Townes at Columbia University in the early 1950s resulted in the invention of the maser, a device that creates a focused microwave beam using stimulated emission (Rosenberg, 2000). Townes later collaborated with Arthur Schawlow of Bell Laboratories on a paper that outlined how stimulated emission might work at the wavelength of visible light. Bell Laboratories received a patent for the resulting invention, the laser, in 1960 (Spetz, 1995).

Medical uses of the laser quickly became apparent, especially after the invention in 1964 of the argon laser, a light source that promoted photocoagulation. Ophthalmologists and dermatologists were the first and most frequent users of lasers, which enabled retinal reattachment, treatment of glaucoma, and the removal of disfiguring port-wine stains. However, other uses of the laser technique proceeded more slowly. Although clinicians recognized that the wavelength, duration, and energy intensity of a laser beam could be manipulated, fundamental questions about the optical, thermal, and physical properties of tissue had to be answered before lasers could be used to treat other clinical conditions. By the 1980s, many of these uncertainties had been overcome, and lasers were soon used in a wide range of clinical specialties, including gynecology, gastroenterology, and cardiology.

Flexible gastrointestinal endoscopy was first developed in the early 1960s with significant academic contributions by physicists van Heel in Holland and Hopkins and Kapany in the United Kingdom (Gelijns and Rosenberg, 1999). Their work, reported simultaneously in *Nature* in 1954, laid out the principles of coherent image transmission for sending images along an aligned bundle of flexible glass fibers (Hopkins and Kapany, 1954; van Heel, 1954). Van Heel presented the concept of the coated glass fiber and the possibility of plastic coatings, which later turned out to be unworkable. Both papers described the conveyance of optical images along a glass fiber, a concept that had been developed earlier. Hopkins and Kapany also elucidated the basic principles of fiber alignment. Upon reading their work in *Nature,* Hirschowitz, a gastroenterologist at the University of Michigan, Peters, an optical physicist, and Curtiss, an undergraduate student, undertook research to develop a workable fiber-optic instrument for visualizing the upper gastrointestinal tract (Gelijns and Rosenberg, 1999). Hirschowitz tested the first operating gastroscope on himself in February 1957. The academic trio subsequently licensed the technology to American Cystoscope Makers, Inc., and collaborated with the firm to develop the first commercial flexible endoscopes.

Basic advances in physics were essential to the development of all imaging technologies (Gelijns and Rosenberg, 1999). These advances were typically generated in departments of physics at universities in Europe and the United States; they can be traced all the way back to Roentgen, a professor of physics at the University of Würzberg in the nineteenth century.

As these examples illustrate, collaboration between AMCs and physical science and engineering departments has been a consistent pattern in the medical device industry. In some cases, collaborations date back many years. Current collaborative efforts often are continuations of long-standing cooperation between engineers, scientists, and medical faculty. Tissue engineering is an example. One of the most productive collaborations was between Joseph Vacanti, a pediatric surgeon at Massachusetts General Hospital in Boston, and Robert Langer, a professor of chemical engineering at MIT. Basically, tissue engineering involves creating a scaffold of an artificial, biodegradable polymer, which is then seeded with living cells and immersed in growth factors. As the cells multiply, they fill up the scaffold and grow into a three-dimensional tissue. R&D was focused on creating organs and body parts, such as bone, skin, pancreas, teeth, breast, heart valves, arteries and veins, urinary tract, and cartilage. This research, and the involvement of academic faculty in the creation of start-up firms, spawned an entirely new industry. Lysaght et al. (1998) documented the creation of 40 start-up firms in tissue engineering, 10 of which had gone public at a market capitalization of $1.7 billion in January 1998.

The increase in interdisciplinary research collaborations has also been stimulated by several federal funding initiatives. For instance, several engineering research centers funded by NSF perform research relevant to medical devices. These include the Center for the Engineering of Living Tissues at Georgia Institute of Technology and the Emory School of Medicine, the Engineering Research Center for Computer-Integrated Surgical Systems and Technology at Johns Hopkins University, the Engineered Biomaterials Engineering Research Center at the University of Washington, and the Engineering Research Center in Bioengineering Educational Technologies at Vanderbilt University. Other agencies have also funded interdisciplinary research centers. The Center for Integration of Medicine and Innovative Technology (CIMIT), established with funding from the Massachusetts General Hospital and DOD, involves various clinical specialties in the Partners Health Care System in Boston, MIT, and Draper Laboratory, as well as industrial partners in the Partners Health Care System in Boston (Parrish, 1998). CIMIT is devoted to the development and evaluation of innovative diagnostic and therapeutic devices.

Case Study: Center for Integration of Medicine and Innovative Technology

CIMIT was founded in 1993 to accelerate the generation, development, and time-to-practice of innovative and high-impact concepts in minimally invasive therapy that improve the quality and lower the cost of health-care delivery. CIMIT operates as a consortium that includes Massachusetts General Hospital and the Brigham and Women's Hospital, both of which provide clinical expertise, and MIT and Draper Laboratory, which provide basic technical and engineering expertise.

In 1998, with the consortium structure in place, CIMIT won a competitive federal award for minimally invasive health-care technologies, administered through the U.S. Army. This unconventional funding source allows CIMIT to undertake high-risk, high-reward research that spans a wide range of scientific and technical fields. Army funding allows CIMIT to pursue short-term, unconventional, developmental, integrative, and sometimes speculative research, in addition to longer term, clinical, and basic science projects.

To maximize the potential for interdisciplinary collaborations, CIMIT is organized as a matrix, with clinical focus areas: cardiovascular; stroke; trauma and critical care, supported by a technological infrastructure comprised of technology teams; biomaterials; endoscopic tools; endovascular tools; energy delivery; medical imaging; microsensors; simulation and modeling; surgical planning; and tissue engineering. CIMIT also has an outcomes and technology assessment program to analyze the cost effectiveness of new therapies and devices. Research activities are supplemented by a broad industry collaboration initiative, a fundraising initiative, and an education/outreach initiative.

Industry collaboration is a major component of CIMIT, which uses a hierarchical mechanism for industry-funded research. CIMIT conducts project-specific research funded or performed in conjunction with industry. At the next level, companies interested in multiple areas of CIMIT research may join the CIMIT industrial partnership program in order to gain access to physician consultation, prepublication reports, and symposiums. Lastly, corporations with a business interest in integrative technologies that would otherwise have to negotiate multiple research agreements with multiple departments can become strategic alliance partners of CIMIT, thereby making a long-term commitment to research in minimally invasive therapy and ancillary technologies. Industry gains early access to opinion leaders, novel technologies, and interdepartmental expertise. For CIMIT, industry collaboration provides a research focus as well as additional research dollars. CIMIT enables the Army to leverage research dollars and provides a conduit for transitioning research and technology into clinical practice or the device industry.

Academic Medical Centers

AMCs conduct research that contributes to the development and diffusion of medical devices. AMCs have been involved in: (1) generating knowledge about human physiology and pathophysiology; (2) developing product ideas, device prototypes, and manufacturing methods; (3) clinical testing and feedback; (4) modifying existing products; and (5) discovering new indications of use.

Physiology and Pathophysiology

AMCs are an important source of knowledge concerning human physiology and pathophysiology. Understanding the electrophysiology of the heart, for example, is critical for designing a pacemaker or implantable defibrillator, as is circulatory physiology for designing artificial hearts and circulatory assist devices. Similarly, knowledge of renal physiology, which has been crucial in elucidating the pharmacodynamics and pharmacokinetics of hemodialysis, has contributed to the development of improved dialysis machines. Research increasingly involves collaborations between basic science departments (e.g., physiology and pharmacology), engineering departments (e.g., mechanical, electrical, and chemical), and clinical departments (e.g., surgery and medicine).

Another important research focus is health information systems. Compared to other service industries, especially financial services, the diffusion of information technology in health care has been slow. The role of university faculty in developing and commercializing health information systems is reviewed in the following case study.

Case Study: Development of the Medical Information System Industry in the United States

The advent of Medicare and Medicaid in the mid-1960s marked the beginning of the transformation of the health-care delivery system. These federal programs provided enormous new resources for medical care, but also initiated significant new reporting requirements for receiving institutions. The combination of increased patient demand and the need for a more sophisticated administrative infrastructure created a demand for information. At the same time, the development and industrial use of information technology developed to the point that it could be adapted to the health-care industry. One of the major barriers to the transfer and growth of information technology was the absence of industrial organizations capable of financing, absorbing, and adapting the technology. In the 1960s, most hospitals were charitable institutions, and most physicians were in solo practices.

Research in medicine (carried out in medical schools and AMCs) has been concentrated in the scientific and practice arenas, focused on basic causes of disease and the drugs, devices, and procedures to treat them. However, with the entry by the federal government into the direct delivery of medical care, Congress provided funds for research into medical computer applications, mainly through NIH, but also through a number of directed programs in other federal organizations and the newly created National Library of Medicine.

By the 1970s, two strategies emerged. The first, derived from the research support of NIH and other agencies, was a new specialty—medical informatics—in a relatively small number of institutions and mainly federally supported. The second, the business needs of hospitals and doctors, spawned the development of a growing number of commercial enterprises, often as outgrowths of IBM, which dominated the field of business applications for hospitals. Information technology rapidly expanded from purely business applications to more clinical areas and, more recently, into decision support areas.

Compared to other service industries, especially financial services, the dispersion of information technology in health care has been slow. In 2001, almost 38 years after Medicare and Medicaid were introduced, *Healthcare Informatics*, which lists the top 100 medical information technology companies by revenue, found only four that had exceeded $1 billion in revenue; the next highest was $523 million, and the next $435 million. Number 100 was $3.7 million. The total market is estimated to be as much as $50 billion (in a health-care industry of more than $1 trillion). The direct, traceable transfer of academically developed systems to commercial use has followed two tracks: (1) direct transfer of results of government and/or institution-sponsored programs; and (2) the results of privately funded programs.

In the late 1960s, government agencies gave grants to academic institutions for R&D on computer systems. In almost every case, funds were matched by internal institutional funds because it was believed that these projects would also be beneficial to the operation of the hospital. After initial development, the hospital either gave or sold the programs to a company to develop systems for other hospitals. The second track was university personnel who left and began start-up companies that have become significant commercial enterprises.

An example of a major academic contribution began with research at the Laboratory of Computer Science at the Massachusetts General Hospital (MGH), one of the principal teaching hospitals of Harvard Medical School. Government-sponsored research resulted in important commercial outcomes, such as the language MUMPS (MGH utility multiprogramming system), which went on to be used in many applications and was supported by both a MUMPS users group and the MUMPS Development Committee, which managed the MUMPS ANSI Standard.

One early example of an institution-sponsored transfer was the work of Homer Warner of the Latter Day Saints (LDS) Hospital, the principal teaching hospital of the University of Utah Medical School (later named Intermountain Health System), in the development of computerized hospital information systems. LDS developed both MEDLAB and Health Evaluation through Logical Processing (HELP). In the 1980s, the rights were acquired by the 3M Company and commercialized as a leading clinical information system.

Another example, which resulted from research funded directly by government that was spun off into a private company, is Public Health Automated Medical Information System (PHAMIS). Initially, a government contract to automate the records of the public health hospitals was given to Malcolm Glaser at the University of Washington. In the 1980s, the government closed the hospitals and transferred ownership of the resulting information system to Glaser who started the commercial company. The PHAMIS hospital information system was named Lastword. The company went public and was acquired by the IDX company in 1997. The combined company was the tenth largest in the health information technology industry in 2001.

A second spin-off of MGH was Meditech, now the twentieth largest health information technology company. The principals who developed MUMPS left MGH to form Meditech and it has developed several generations of proprietary languages following MUMPS (MIS and MAGIC) and used them to develop hospital information systems that are now installed in more than one thousand hospitals.

A final example relates to two efforts at the Beth Israel and Peter Bent Brigham Hospitals (PBBH), also principal teaching hospitals of Harvard Medical School. A program begun in 1976 under the leadership of Warner Slack and Howard Bleich, both professors at Harvard Medical School, sponsored mainly by government grants, expanded from research on medical informatics to operational systems for laboratories, pharmacies, other laboratories, and routine services. A second project (known as BICS) at PBBH focused on order entry and other operational functions. Commercial spin-offs resulted from both projects.

A related example involves the work of Dr. Dennis Gillings, a statistician and epidemiologist from the University of North Carolina School of Public Health. Dr. Billings started a company that grew into Quintiles, a major corporation that supports clinical trials, pharmacoeconomics, and the health service research needs of the pharmaceutical, biotechnology, and medical device industries.

Examples of privately funded research that have contributed to the industry also abound. One example of the direct translation of an AMC product development project funded by private dollars to a public company is Transition Systems, Inc. The system was originally developed as an internal clinical cost-accounting system for the growing managed care market. New England Medical Center received private foundation support in 1981 to develop a management control system that combined clinical and financial data. When Medicare shifted from cost-based reimbursement to prospective payments to providers based on diagnosis-related groups (DRGs) in 1984, there was a widespread demand for cost-accounting systems, and New England Medical Center transformed the project team into a corporate spin-off, retaining the majority ownership. The company prospered and went public in conjunction with an investment group partnership 10 years later.

A second example is a company whose products were developed by the Health Policy Institute at Boston University. Health Payment Review was formed with venture capital funding to market products that added clinical appropriateness to the payment methods of managed care companies. After going public, the company was acquired by HBO, Inc., which, in turn, was acquired by McKesson in 1999, the third largest health information technology company in 2001.

The flow of researchers forming new commercial start-up companies continues. The managers of many companies that serve niche markets, such as electronic medical records and disease management, started as participants in academic research teams. Thus, the array of information technology, which is becoming critical to the success of managed care, is likely to be infused with a continuing stream of developments from AMCs to industry. However, as AMCs become more insistent on retaining value from the results of their research, they are beginning to take equity positions in companies willing to commercialize their research. For example, the PBBH's BICS system has been incorporated into the public company, Eclypsis, in which Partners Health Care System, the parent of the PBBH, holds an equity position.

In conclusion, it appears that the greatest impact of AMC research on information technology development has come from entrepreneurs who leave academic environments with research experience and ideas of how that experience can be transformed into commercial products. Successful individuals have garnered venture capital and eventually either consolidated companies into larger health information companies or made public offerings as independent companies.

Ideas, Prototypes, and Manufacturing Methods

Clinicians/academic researchers not only identify the need for new devices or improvements in existing devices, but, because they are also the eventual users of their devices, they are often the innovators and builders of original prototypes. Von Hippel and Finkelstein (1979), for example, described how users were involved in the invention of the automated clinical chemical analyzer. Other studies in the area of renal dialysis, intrauterine devices, catheters, and MRI machines have presented similar findings (Shaw, 1987, Gelijns 1991, Gelijns and Rosenberg, 1995; Gelijns et al., 1998). However, academic researchers are often unable to advance projects beyond a certain point because critical enabling technologies are missing or are too specialized to be developed in the laboratory or elsewhere in the university. To overcome this hurdle, researchers often form partnerships with industrial firms with the applicable technological expertise and interest in the proposed application.

The contributions of academic faculty in the development of device prototypes can be documented for the whole clinical spectrum of medical device categories, ranging from diagnostic devices to therapeutic devices. In diagnostic devices, for example, Robert Ledley, professor of physiology, biophysics, and radiology at Georgetown University Medical Center, developed the first prototype for a whole body CT scanner, patented the device, and created a company, Digital Information Systems, to commercialize it. Pfizer ultimately licensed the device and introduced it into clinical practice in 1975.

In therapeutic devices, numerous examples show the central role of clinical faculty in the development of new or modified devices. In gynecological laparoscopy, for example, the gynecologist Kurt Semm at the University of Kiel worked with the device company Storz to develop a whole range of instruments that could be moved through the operative channel of the endoscope or any other cannula (Gelijns and Rosenberg, 1999). Through Storz's close collaborations with Semm and Hopkins, this firm became the leading manufacturer of gynecological laparoscopes worldwide. The cardiologist Andreas Gruentzig, then at the University of Zurich, collaborated with the device firm Schneider to develop the first percutaneous, transluminal, coronary angioplasty catheter. The role of AMCs in the development of focused ultrasound therapies is another case in point.

Case Study: Development of Focused Ultrasound Therapies

Clinical ultrasound works much the same way as radar—energy is produced by a transducer, and the reflected energy is received and processed by a receiver. The time between signal transmission and reception correlates directly with distance, and the amplitude of the return indicates the material properties of the reflecting surface. When an array of acoustic transducers is used, the resulting fan-beam image shows distance and amplitude as a function of the placement of the transducer along the array. In this mode, ultrasound is used primarily for diagnostic purposes, such as cardiovascular or fetal imaging. Just as optical lenses can focus light on a single spot, however, acoustic lenses can focus ultrasound on a single spot. In this mode, ultrasound becomes a therapeutic tool, rather than a strictly diagnostic tool.

The principles behind using acoustic energy for diagnostic purposes have been known since at least 1942, when researchers at Columbia University demonstrated the operation of a focused ultrasound generator capable of producing focal heating in paraffin blocks, liver tissue, and inside the brains of animals (Lynn et al., 1942). Subsequently, significant advances were made at the University of Illinois, as well as the Massachusetts General Hospital and Harvard Medical School. These early applications of focused ultrasound were used to examine central nervous system tissue (Fry et al., 1955) and the brain (Basauri and Lele, 1962; Fry and Fry, 1960; Lele, 1962). By the late 1970s, the broader use of ultrasound in surgery was considered a viable treatment modality (Fry, 1978; Lele, 1975).

By the 1970s, the dominant obstacle to the use of focused ultrasound for therapy was no longer the delivery of acoustic energy, but the inability to monitor the extent of the therapy. There are two aspects to this obstacle. First, because focused ultrasound must operate with a relatively small focus spot to deliver sufficiently high energy, real-time monitoring of the focus location is necessary to ensure that the entire target zone has been treated. Monitoring is particularly important to determine the duration of the therapy—sometimes up to several hours. Second, because the purpose of focused ultrasound therapy is to induce either coagulation or tissue necrosis, the inability to monitor the induced temperature changes in the targeted tissue meant that it was nearly impossible to establish rigorous treatment protocols. For these reasons, advances in magnetic resonance (MR) physics, which led to an understanding of how MR can be used to monitor temperature, and the integration of magnetic resonance imaging (MRI) with focused ultrasound therapy for image-based guidance became key enabling technologies for using focused ultrasound for therapy. Based on the interrelationship between focused ultrasound therapy and MR, companies with significant business interests in high-end imaging equipment, such as GE, began to invest in sponsored research in focused ultrasound. GE sponsored the Brigham and Women's Hospital—an AMC affiliated with Harvard Medical School—to further develop this technology. Indeed, the collaboration between Brigham and GE was so complete that GE researchers were contributing, or even leading, authors of several of the major papers on the topic of focused ultrasound (e.g., Cline et al., 1992, 1994).

By the early 1990s, the potential for focused ultrasound was beginning to be realized. Numerous studies on the use of focused ultrasound for prostate hyperplasia were conducted at various AMCs; focused ultrasound therapy on brain tissue without prior removal of a section of the skull was initially demonstrated at Brigham (Hynynen and Jolesz, 1998), as well as new treatment options for the ablation of breast fibroadenomas (Hynynen et al., 2001). As of the late 1990s, numerous research activities in focused ultrasound were under way at the University of Michigan, the Mayo Clinic, and elsewhere (Spera, 1998).

Although the market demand for focused ultrasound technology remains low, many clearly believe in its potential. Focus Surgery, Inc., for example, has secured licenses for therapeutic applications of focused ultrasound in a number of organ systems, including the prostate, brain, liver, kidney, pancreas, and breast. MRI manufacturers, including GE, Siemens, and Phillips, are all believed to be actively pursuing this technology (Spera, 1998). Like other imaging technologies (e.g., CT, MR), focused ultrasound is a technology that was born in academic research settings and has gained commercial interest. Industry is now turning to AMCs for both new insights and clinical validations.

Clinical Testing and Feedback

Improvements in product design depend in large measure on extended clinical testing that requires close collaborative relationships between industry and academia, sometimes involving several major medical schools and their teaching hospitals (Gelijns and Rosenberg, 1999). The clinical data generated by testing not only provide feedback for altering product designs, but also provide a basis for obtaining FDA premarketing and payer coverage approval, and thereby lead to widespread market access. In recent years, spending on clinical trials by industry has increased substantially.

AMCs have traditionally been involved in the testing of prototype devices and have been the source of patients for extensive clinical trials. AMCs have been the venue of care for patients who need implantable devices and invasive procedures. AMC faculty members were often involved in designing, conducting, and analyzing clinical trials, but in the last decade contract research organizations

(CROs) have captured part of this market (Moskowitz and Thompson, 1997). CROs are private, for-profit organizations engaged in the management of clinical trials, including protocol design, patient recruitment, data collection, data management, monitoring, and analysis. In the medical device industry, CRO use is not common; only 13 percent of medical device firms use CROs (whereas 90 percent of drug firms use CROs) (Centerwatch, 2001).

Only a small percentage of devices (i.e., Class 3 FDA devices and a small subset of 510(k) devices) must undergo rigorous safety and efficacy evaluation. As a result, the overall number of randomized controlled trials for devices is low. The number is slowly increasing, however, as the FDA grapples with changing its policies about which devices require rigorous evaluation. The FDA must ensure that device trials take into account ethical, technical, and methodological challenges at various stages of the evaluation process.

To begin with, choosing the optimal time to initiate a device trial is more of a challenge than the same decision for a pharmaceutical. A pharmaceutical compound generally does not undergo substantial changes as it progresses from animal to human studies. Devices, however, undergo extensive modifications and refinements during the development phase, and early evaluations run the risk of failure or, at least, the need for redesign and retesting, which entails time and monetary expenses that few start-up companies can afford.

Once the optimal time to begin a clinical evaluation is established, decisions concerning which venue and which clinicians to engage to test the device can have a major effect on how the results of the trial will be interpreted and whether the device is widely adopted. In contrast to pharmaceuticals, the efficacy of a surgically implanted device can be linked to the skill of the implanting surgeon. Thus, conducting a trial in a highly specialized medical center that has unique surgical expertise may result in a successful trial but may not lead to widespread use.

Blind studies, an important technique for controlling observational bias in evaluating the safety and efficacy of new clinical interventions, is also problematic in trials of invasive or implantable devices. The clinician who implants a device cannot implant placebos; blind studies are not possible when the comparative therapy is not a device. Randomization is also a problem in device trials, especially in the case of a life-threatening illness, in which case both patient and physician expect that the device is their best hope and would be devastated to learn, up front, that they would not receive the preferred therapy. This deters some patients and physicians from entering into device trials; others might enroll but seek treatment outside the protocol if they don't receive the therapy they want. This might lead to a loss-to-follow-up or out-of-protocol crossover, which could ruin a small-scale trial. The ethical dilemma is heightened when there are no alternative therapies and assignment to a control group means essentially no therapy (Moskowitz and Thompson, 1997).

Measuring survival in trials that compare devices and medical therapies poses methodological challenges. When device therapy involves a high up-front operative risk, but subsequently a reduced mortality compared to the control therapy, the survival curves are likely to cross. Analyzing the differences depends on the analytical method chosen and the time frame of the analysis. Most analytical methods (e.g., log-rank, Wilcoxon test) average risk over the follow-up period. So, extending or reducing the follow-up time can potentially reverse the ordering of relative efficacy because less or more weight, respectively, will be given to mortality in the perioperative period (Rose et al., 1999).

Another technical constraint is the limitation of patient recruitment. Devices often have small numbers of potential users and, therefore, few eligible candidates for clinical trials. Also, device implantation and monitoring usually require specialized training or skills that may not be available in large enough numbers to conduct trials at several AMCs.

University research could make significant contributions to evaluative research for medical devices by addressing some of the methodological challenges of device randomized control trials (RCTs). Moreover, academic analysis could clarify the bases for policy changes at the FDA, for example, with respect to the strength and limitations of RCTs, and the implications of expediting the approval process (e.g., shifting some of the premarketing research to the postmarketing setting).

Product Modification and Discovery of New Indications of Use

Of course, the development process does not end with the widespread introduction of new products into practice. Their eventual uses depend on an extensive improvement process that vastly increases their practical applications. Users, often clinicians in AMCs, provide necessary feedback about the shortcomings of new devices. Consider, for instance, the evolution of endoscopes. Today's "cold-light" video-endoscope, with a computer-chip camera at its tip that can be used both for diagnosis and therapy, is a world apart from its predecessor in the 1950s. During those years, for instance, the lamp at the tip of the endoscope could cause serious burns, vision was often restricted, the quality of images was poor, therapeutic applications were essentially nonexistent, and obtaining permanent documentation of the images was highly problematic. Feedback from users encouraged manufacturers to develop subsequent generations of endoscopes. Whereas the evolution of endoscopic technology did indeed require a few major improvements, such as the introduction of fiber optics and video capabilities, its current characteristics are the result of a continuous flow of refinements that have resulted in increased flexibility, miniaturization, and improved visibility, which have vastly expanded the therapeutic possibilities of endoscopy (Gelijns and Rosenberg, 1999).

In addition, clinicians can expand the applications of a device to new clinical uses. In the case of GI endoscopy, for instance, academicians expanded the use of fiber-optic endoscopes from the upper GI tract to gastroenterological areas, such as the esophagus, duodenum, and colon. Lasers were originally introduced for ophthalmologic and dermatologic purposes but are currently being used for a wide variety of indications in gynecology, cardiac surgery, and oncology, to name but a few. The identification of new applications, sometimes totally unexpected, is an important contribution of academic researchers to the medical device industry (Rosenberg, 1996; Gelijns et al., 1998).

Mechanisms of Transfer from Academia to Industry

Advances have been transferred to industry by various routes. Traditionally, research advances were placed in the public domain either through publications or presentations at conferences. Another common practice was to hire academic researchers as consultants or researchers, sometimes after firms had sponsored their research. Another pathway that has expanded very rapidly in recent decades is university patenting and licensing practices. Pfizer licensed Georgetown's whole body scanner, whereas Syntex, Varian, and GE all entered the CT field by licensing important technical improvements from research at Stanford University. University faculty members have also been active in the creation of start-up firms to develop and market their inventions.

IMPACT ON INDUSTRIAL PERFORMANCE

Past and Present Contributions

One of the defining characteristics of the medical devices and equipment sector is a strong dependency between universities and industry. Based on the results of its fact-finding efforts, the panel concludes that academic research has had a substantial impact on the industry's performance. The contribution of universities goes well beyond educating new generations of employees and making fundamental advances in scientific and technological knowledge that may contribute to the development of new medical devices. It includes a high degree of involvement in product development, product evaluation and introduction, and product modification.

In making this observation, the panel does not wish to diminish the long-term importance of training people in research techniques or making fundamental advances in the scientific and technological knowledge base. In fact, basic advances in physics, mathematics, and chemistry have directly contributed to a whole range of medical devices and equipment. Moreover, with the integration of the biological sciences and the engineering sciences, as in tissue engineering, the contributions of university research may be even greater in the future. Nevertheless, the panel wants to highlight the role of AMCs in the development, clinical testing and evaluation, modification, and extensions of use of prototype devices. The case studies above represent just a few illustrations of their importance.

Trends, Opportunities, Challenges, and Gaps

Although university research has made substantial contributions to the medical devices and equipment industry in the past, the rapidly changing health-care environment is creating both new opportunities and new challenges for university-industry interactions. In recent years, the NIH budget has grown, allowing for an increase in research on the biological bases of health and disease. The increase in NIH funding does not, however, obviate the need to address questions about the allocation of these funds for different types of research. Traditionally, NIH support for research closely coupled to the development of medical devices has been limited. Recently, NIH, as well as NSF, created new initiatives to encourage bioengineering research to compensate for the planned closing in 2006 of the Whitaker Foundation, which has provided significant support for bioengineering research in the past (Whitaker Foundation, 2001). Moreover, most federal investment in biomedical research goes to support laboratory-oriented (or nonhuman subjects) research; much less support is allocated for studies of the very diverse activities that come under the rubric of clinical research. The latter have traditionally depended heavily on internal funding from academic health centers, particularly cross-subsidies from patient care revenues. As pressures for cost containment increase and clinical faculty compete for contracts with managed care organizations, however, clinical income has decreased substantially, which means less money is available to cross-subsidize research. In addition, recent studies have shown that academic faculty in regions with high managed care penetration publish fewer papers and are less likely to be awarded NIH grants (Griner and Blumenthal, 1998). Thus, although NIH funds continue to increase, changes in the financing of medical care are creating serious uncertainties in the funding flow that sustains clinical research.

Various disciplines and various schools in universities contribute to the development of medical devices. However, establishing interdisciplinary links in the university between faculty in the natural sciences and engineering with faculty in medicine has been difficult in the past. With the emergence of new fields of research, such as tissue engineering, and encouraged by interdisciplinary research funding initiatives by both NIH and NSF, creating interdisciplinary linkages may be easier in the future.

Although there has been a strong interdependency between universities and medical devices and equipment firms in the past, opportunities are being created for more systematic interactions. With the rapid increase in the costs of conducting research, for example, universities and companies may look for ways to share basic facilities, such as animal laboratories or expensive equipment (e.g., a proton beam unit) (Their, 1998). Another mechanism for improving university-industry relations may be the creation of more systematic research partnerships. One interesting model is the Center for Innovative Minimally Invasive Therapy, which involves faculty from AMCs and the physical and engineering sciences, as well as industry partners (Parrish, 1998).

Systematic partnerships may also have considerable payoffs in product modification and the discovery of new indications of use. For most medical devices, new uses result from application to other organ systems, although these transfers often require design modifications. The first endoscopes, for example, were used for cystoscopy early in this century. In the 1960s, after the development and introduction of fiber optics, GI endoscopy and gynecological laparoscopy became well established. It took nearly four decades to transfer laparoscopy from gynecology to general surgery, where it transformed gallbladder surgery. Earlier identification of such secondary indications may have substantial benefits, for society and for industry.

Universities are an important location for clinical testing. Universities, as well as private CROs in recent years, have been active in designing, conducting, and analyzing clinical trials. Major questions, however, remain about the appropriate evaluation of new devices, especially innovative and implantable devices. These questions differ significantly from questions about the evaluation of pharmaceuticals.

Traditionally, the research results of AMCs that have been most important in the development of medical devices were not patented but were placed in the public domain through open publication. In recent years, as a result of a number of changes in federal policy, there has been a major upsurge in university patenting and licensing. The panel has little doubt that this increase in patenting has strengthened university-industry interactions, to the benefit of both the economy and the university. Despite these benefits, however, the panel believes some hard thinking and empirical research should be done to assess the consequences of these changes on the role of universities in the innovation system. Have these developments indeed increased the effectiveness of technology transfer from universities to industry? Or would the licensed technologies have been picked up by industry anyhow? And what are the unintended consequences? Are universities changing the nature of their research activities from fundamental, long-range research to applied research? Has the upsurge in university patenting increased the transaction costs of science? Are universities licensing inventions that can be classified exclusively as research tools? All of these questions should be addressed.

RECOMMENDATIONS

The panel was asked to examine the contributions of academic research to the medical devices and equipent industry and to delineate ways of improving such contributions in the future. This report provides evidence that academic research has contributed strongly to industrial performance in the medical devices and equipment sector; at the same time, steps can be undertaken to improve these contributions.

Recommendation 3-1. The panel concurs with recent recommendations by the Commonwealth Task Force on Academic Health Centers that the National Institutes of Health and other institutions should recognize the importance and vulnerability of clinical research by increasing support for clinical research at academic medical centers.

Recommendation 3-2. Optimizing the contributions of university research will require creating effective linkages between faculty in engineering schools and faculty in medicine. The panel recommends that universities invest in interdisciplinary centers to generate new knowledge for advancing medical devices and to develop new diagnostic and therapeutic modalities. Universities are also encouraged to decrease barriers to conducting interdisciplinary research. Funding agencies should carefully evaluate new interdisciplinary programs and initiatives in biology/medicine and engineering and encourage the growth of the most promising ones.

Recommendation 3-3. Universities and medical device firms should explore ways of creating more systematic partnerships between universities (especially academic medical centers) and industrial firms for the development and evaluation of new, cost-effective medical devices. Models worth contemplating include interdisciplinary centers for the development and evaluation of medical devices that include industrial partners, the sharing of expensive facilities (e.g., animal laboratories), exchange fellowships, and the teaching of joint courses. Moreover, the panel believes that both society and the medical device industry would benefit substantially if new indications of use could be identified sooner after the development of a device. To expedite the discovery of new indications, device manufacturers might draw more fully on interdisciplinary panels of academic experts who would consider how a new technological capability (e.g., lasers or positron emission tomography) that is useful for one purpose might also be useful (modified as necessary) in another field.

Recommendation 3-4. Federal agencies that fund academic research relevant to the medical device industry should support research on the effectiveness of current incentives for transferring research findings to the industry and ways of improving the transfer process. Given the short product life cycles of many medical devices, the timing of decisions and processes pertaining to transfer affects the short windows of commercial opportunity.

Recommendation 3-5. Academic researchers should bring together industry, regulatory, and clinical panels to discuss requirements for device evaluations. Discussions should include regulatory requirements (e.g., market clearance by the Food and Drug Administration), third-party payment eligibility, market research, and information dissemination/marketing issues (e.g., direct-to-consumer advertising). Regulation, payment/reimbursement systems, and marketing all have profound effects on the pathway for getting device concepts to users. Therefore, anticipating and understanding regulatory, payment, and marketing needs should be incorporated and fed back into device design and refinement. Academic centers (including business schools) and industry can share considerable insight and expertise in all of these areas.

Recommendation 3-6. Given that all parties—physicians, patients, manufacturers, and payers—benefit from the rigorous information on the value of new and improved medical devices, the panel recommends that payers, National Institutes of Health, and medical device firms define the circumstances under which public-private support for device trials is appropriate.

REFERENCES

AdvaMed (Advanced Medical Technology Association). 2001. U.S. Medical Technology Industry Statistics. Washington, D.C.: AdvaMed.

Agnew, B. 1998. Multidisciplinary research: NIH plans bioengineering initiative. Science 280(5369): 1516–1518.

Basauri, L., and P.P. Lele. 1962. A simple method for production of trackless focal lesions with focused ultrasound: statistical evaluation of the effects of irradiation on the central nervous system of the cat. Journal of Physiology 160(3): 513–534.

Centerwatch. 2001. Assessing change in CRO usage practices. Centerwatch Newsletter 8(1): Article 201.

Chronicle Information Resources. 1999. The Chronicle of Cardiovascular and Internal Medicine. Mississauga, Ontario: Chronicle Information Resources Ltd.

Cline, H.E., K. Hynynen, C.J. Hardy, R.D. Watkins, J.F. Schenck, and F.A. Jolesz. 1994. MR temperature mapping of focused ultrasound surgery. Magnetic Resonance in Medicine 31(6): 628–636.

Cline, H.E., J.F. Schenck, K. Hynynen, R.D. Watkins, S.P. Souza, and F.A. Jolesz. 1992. MR-guided focused ultrasound surgery. Journal of Computer Assisted Tomography 16(6): 956–965.

Commonwealth Fund Task Force on Academic Health Centers. 1999. From Bench to Bedside: Preserving the Research Mission of Academic Health Centers. Boston, Mass.: The Commonwealth Fund.

Director's Panel on Clinical Research. 1997. Report to the Advisory Committee to the NIH Director. Bethesda, Md.: National Institutes of Health.

Fry, F.J., ed. 1978. Ultrasound: Its Applications in Medicine and Biology. Part I. New York: Elsevier Press.

Fry, W.J., J.W. Barnard, F.J. Fry, R.F. Krumins, and J.F. Brennan. 1955. Ultrasonic lesions in the mammalian central nervous system. Science 122(3168): 517–518.

Fry, W.J., and F.J. Fry. 1960. Fundamental neurological research and human neurosurgery using intense ultrasound. IRE Transactions on Medical Electronics 7: 166–181.

Gelijns, A.C. 1991. Innovation in Clinical Practice: The Dynamics of Medical Technology Development. Washington, D.C.: National Academy Press.

Gelijns, A.C., and N. Rosenberg. 1995. From the Scalpel to the Scope: Endoscopic Innovations in Gastroenterology, Gynecology, and Surgery. Pp. 67–96 in Medical Innovation at the Crossroads. Volume VI. Sources of Medical Technology: Universities and Industry, A.C. Gelijns and N. Rosenberg, eds. Washington, D.C.: National Academy Press.

Gelijns, A.C., and N. Rosenberg. 1999. Diagnostic Devices: An Analysis of Comparative Advantages. Pp. 312–358 in The Sources of Industrial Leadership in Seven Industries, D. Mowery and R. Nelson, eds. Boston: Cambridge University Press.

Gelijns, A.C., N. Rosenberg, and G. Laubach. 1994. Medical Device Innovation: Opportunities and Barriers for Small Firms. Washington, D.C.: National Academy Press.

Gelijns, A.C., N. Rosenberg, and A.J. Moskowitz. 1998. Capturing the unexpected benefits of medical research. New England Journal of Medicine 339(10): 693–698.

Goodnight, J. 2002. Statement Submitted to the Record, Committee on Small Business, U.S. House of Representatives. June 21, 2001. Available online at: *www.hhs.gov/asl/testify/t010621.html.* [June 24, 2003]

Griner, P., and D. Blumenthal. 1998. Reforming the structure and management of academic medical centers: case studies of ten institutions. Academic Medicine 73: 817–825.

Healthcare Informatics. 2001. The Healthcare Informatics 100. Available online at: *http://www.healthcare-informatics.com/issues/2001/06_01/jun01.htm.* [June 24, 2003]

Herman, W.A., D.E. Marlowe, and H. Rudolph. 1998. Future Trends in Medical Device Technology: Results of an Expert Survey. Rockville, Md: Center for Devices and Radiological Health, U.S. Food and Drug Administration. Available online at: *http://www.fda.gov/cdrh/ost/trends/TOC.html.* [June 24, 2003]

Hopkins, H.H., and N.S. Kapany. 1954. A flexible fibrescope, using static scanning. Nature 173: 39–41.

Hynynen, K., and F.A. Jolesz. 1998. Demonstration of potential noninvasive ultrasound brain therapy through an intact skull. Ultrasound in Medicine and Biology 24(2): 275–283.

Hynynen, K., O. Pomeroy, D.N. Smith, P.E. Huber, N.J. McDannold, J. Kettenbach, J. Baum, S. Singer, and F.A. Jolesz. 2001. MR imaging-guided focused ultrasound surgery of fibroadenomas in the breast: a feasibility study. Radiology 219(1): 176–185.

Iglehart, J.K. 1994. Rapid changes for academic medical centers. Part 1. New England Journal of Medicine 331(20): 1391–1395.

Keller, K.H. 1998. How are Changes in the Health Care Environment Affecting University-Industry Research Collaboration? Presentation at the NAE workshop on Medical Devices and the University-Industry Connection: Future Directions, Washington, D.C., November 2, 1998.

Lele, P.P. 1962. A simple method for production of trackless focal lesions with focused ultrasound. Journal of Physiology 160: 494–512.

Lele, P.P. 1975. Ultrasound in surgery. Pp. 325–340 in Fundamental and Applied Aspects of Nonionizing Radiation, S.M. Michaelson, E.L. Carstensen, R, Magin, and M.W. Miller, eds. New York: Plenum Press.

Lewin Group. 2000. Outlook for Medical Technology Innovation. Washington, D.C.: Health Industry Manufacturers Association.

Littell, C.L. 1994. Datawatch. Innovation in medical technology: reading the indicators. Health Affairs 13(3): 226–235.

Lynn, J.G., R.L. Zwemer, A.J. Chick, and A.E. Miller. 1942. A new method for the generation and use of focused ultrasound in experimental biology. Journal of General Physiology 26(1): 179–193.

Lysaght, M.J., N.A. Nguy, and K. Sullivan. 1998. An economic survey of the emerging tissue engineering industry. Tissue Engineering 4(3): 231–238.

McGraw-Hill and U.S. Department of Commerce. 2000. U.S. Industry and Trade Outlook '00. New York: McGraw-Hill.

Moskowitz, J., and J.N. Thompson. 1997. Preventing the extinction of the clinical research ecosystem. Journal of the American Medical Association 278(1): 241–245.

NSF (National Science Foundation). 2000. Federal R&D Funding by Budget Function: Fiscal Years 1998–2000. Arlington, Va.: NSF.

NSF. 2001. Federal Funds for Research and Development: Fiscal Years 1999, 2000, 2001. Arlington, Va.: NSF.

OTA (Office of Technology Assessment). 1984. Federal Policies and the Medical Devices Industry. Washington, D.C.: Government Printing Office.

Parrish, J. 1998. Why CIMIT? Why Now? Research Perspectives Online. Available online at: *http://www.mgh.harvard.edu/DEPTS/pubaffairs/ResearchPerspectives/Aug98whyCIMIT.htm.* [June 24, 2003]

Rose, E.A., A.J. Moskowitz, M. Packer, J.A. Sollano, D.L. Williams, A.R. Tierney, D.F. Heitjan, P. Meier, D.D. Ascheim, R.G. Levitan, A.D. Weinberg, L.W. Stevenson, P.A. Shapiro, R.M. Lazar, J.T. Watson, D.J. Goldstein, and A.C. Gelijns. 1999. The REMATCH trial: rationale, design, and end points: randomized evaluation of mechanical assistance for the treatment of congestive heart failure. Annals of Thoracic Surgery 67(3): 723–730.

Rosenberg, N. 1996. Uncertainty and Technological Change. Pp. 334–356 in the Mosaic of Economic Growth, R. Landau, T. Taylor, and G. Wright, eds. Palo Alto, Calif.: Stanford University Press.

Rosenberg, N. 2000. Interfaces between Universities and Industry. Submitted to the Canadian Center for Advanced Studies.

Shaw, B.F. 1987. The Role of the Interaction between the Manufacturer and the User in the Technological Innovation Process. Ph.D. dissertation. University of Sussex, Sussex, United Kingdom.

Spera, G. 1998. The next wave in minimally invasive surgery. Medical Device and Diagnostic Industry 20(8): 36–44.

Spetz, J. 1995. Physicians and Physicists: The Interdisciplinary Introduction of the Laser to Medicine. Pp. 41–66 in Sources of Medical Technology: Universities and Industry, N. Rosenberg, A. Gelijns, and H. Dawkins, eds. Washington, D.C.: National Academy Press.

Thier, S. 1998. Academic Medicine and the Development of Prototype Technology. Presentation at the NAE workshop on Medical Devices and the University-Industry Connection: Future Directions, Washington, D.C., November 2, 1998.

U.S. Bureau of the Census. 2001. Statistics for Industrial Groups and Industries, 1999. Annual Survey of Manufacturers. Washington, D.C.: U.S. Government Printing Office.

van Heel, A.C.S. 1954. A new method of transporting optical images without aberrations. Nature 173: 39.

Von Hippel, E.A., and S.N. Finkelstein. 1979. Analysis of innovation in automated clinical chemistry analyzers. Science and Public Policy 6(1): 24–37.

Watson, J.T., et.al. 1994. NHLBI Program History. Pp. 27–32 in Report of the Workshop on the Artificial Heart: Planning for Evolving Technologies. Bethesda, Md.: National Institutes of Health.

Whitaker Foundation. 2001. Grant Programs to Phase Out. Available online at: *http://www.whitaker.org/news/phaseout.html.* [June 24, 2003]

Wilkerson Group. 1995. Forces Reshaping the Performance and Contribution of the U.S. Medical Device Industry. Washington, D.C.: Health Industry Manufacturers Association.

Zemel, T. 2002. Venture Capital Investment Stabilizes in 4Q01, According to VentureOne. Press Release, January 28, 2002. San Francisco: VentureOne.

ADDENDUM

E-mail Questionnaire

The following questionnaire was sent to selected individuals in various parts of the medical devices and equipment industry, some of whom attended the November 1998 workshop. Included among the respondents were senior executives at Biomet, Inc., the Center for Integration of Medicine and Innovative Technologies, General Electric Company, Health Quality, IBM, Johnson & Johnson, MedInTec, Inc., Pfizer, and RAND Corporation; professors with expertise in biomedical engineering, mechanical engineering, medical innovation management, and policy from Draper Laboratories, Massachusetts Institute of Technology, and Washington University; and a representative of the Food and Drug Administration.

THE IMPACT OF ACADEMIC RESEARCH ON INDUSTRIAL PERFORMANCE
Medical Devices Panel

We invite your responses to the questions that follow. In addition, please feel free to add any general comments or responses under Question 11 below. Your responses will be used by our Panel as background information for our report. Any material used verbatim will not be attributed to you without seeking your permission.

1. Could you describe briefly significant academic (i.e., university-based research — basic, applied, clinical, etc.) research contributions to the medical devices and equipment industry? (If possible, please supply references to published information that outlines the contributions.)

2. Overall, would you describe the impact of academic research on industrial performance in the medical devices and equipment industry as (Please put an X in one box):

- ☐ 1. very large
- ☐ 2. large
- ☐ 3. medium
- ☐ 4. small
- ☐ 5. very small/non-existent

3. What is the role of academic *research* in educating people who work in your industry? (Please focus on university research activities, rather than university education generally.)

4. What structural forms of university-industry collaboration lead to good results in your industry? An example of such a structure might be a discipline- or industry-oriented "center" that solicits industry sponsors for a collection of projects that span a varied research program, or an academic medical center that provides a venue for clinical research. What seem to be the essential determinants of success of such structures?

5. What are significant emerging trends or problems that the medical devices and equipment industry will face in the future that could benefit from academic research?

6. What changes are required, if any, in academic research if it is to be responsive to these industrial trends and problems?

7. What single step could be taken by universities to enhance the impact of academic research on the industry?

8. What single step could be taken by companies to enhance the impact of academic research on industry?

9. What single step could be taken by government to enhance the impact of academic research on industry?

10. Do you see any downside to enhanced university-industry research collaboration? Things to be avoided?

11. Other comments? Any comments, pointers to other studies, or suggestions would be appreciated.

Workshop Agenda

MEDICAL DEVICES AND THE UNIVERSITY-INDUSTRY CONNECTION: FUTURE DIRECTIONS

November 2, 1998

National Academies Building
2101 Constitution Avenue .NW.
Washington, D.C.

9:00 a.m. **Welcoming Remarks and Overview of the Broader NAE Project**
Jerome Grossman, President, HealthQuality, Inc.

9:15 a.m. **Overview of the Work of the Medical Devices and Equipment Panel**
Annetine Gelijns, (Panel Chair), Director, International Center for Health Outcomes and Innovation Research, Columbia Presbyterian Medical Center

9:40 a.m. **How are Changes in the Health Care Environment Affecting University-Industry Research Collaboration?**
Kenneth Keller, University of Minnesota

10:15 Break

10:30 a.m. **Session I. Basic Academic Scientific and Engineering Research: Contributions to the Medical Device Industry**
Moderator: *Clifford Goodman, The Lewin Group*
Speaker: *Donald Engelman, Yale University*
Speaker: *Robert Nerem, Georgia Institute of Technology*
Respondent: *John Linehan, The Whitaker Foundation*

12 p.m. Lunch in Meeting Room

12:45 p.m. **Session II. Academic Medicine and the Development of Prototype Technology**
Moderator: *Nathan Rosenberg, Stanford University*
Speaker: *Samuel Thier, Partners Health Care*
Respondent: *Paul Citron, Medtronic, Inc.*

2:15 p.m. Break

2:30 p.m. **Session III. Clinical Evaluative Research on Medical Devices: University-Industry Interactions**
Moderator: *Frederick Telling, Pfizer, Inc.*
Speaker: *Richard Rettig, RAND*
Speaker: *Alan Moskowitz, Columbia Presbyterian Medical Center*

4:00 p.m. **Open Discussion. What have we learned today about the impact of academic research on performance in the medical device industry? How can the university research contribution and impact be enhanced?**

4:45 p.m. **Closing Remarks**
Jerome Grossman

Workshop Attendees

Annetine Gelijns, *chair* *
Director, International Center for Health Outcomes and Innovation Research
Columbia Medical Center

James Benson
HIMA

S. Morry Blumenfeld
General Manager, Global Advanced Technology
GE Medical Systems

Paul Citron *
Vice President, Science and Technology
Medtronic, Inc.

Diane Davies
Pfizer

Donald M. Engelman *
Professor, Department of Molecular Biophysics and Biochemistry
Yale University

Marilyn Field
Institute of Medicine

Robert Fischell
MedIntec

Clifford Goodman
The Lewin Group

Jeanne Griffith
Director, Science Resources Studies Division
National Science Foundation

Jerome H. Grossman *
President and CEO
Health Quality Inc.

Elizabeth Jacobson
Deputy Director for Science
Center for Devices and Radiological Health

Kenneth H. Keller
HHH Institute for Public Affairs

John Linehan
The Whitaker Foundation

Stephen Merrill
National Research Council

Dane Miller
Airport Industrial Park
Warsaw, Indiana

Alan Moskowitz
Columbia Medical Center

Richard Nelson
Columbia University

Robert M. Nerem
School of Mechanical Engineering
Georgia Institute of Technology

---

*Panel member

John Parrish
Center for Innovative Minimally Invasive Therapy
Massachusetts General Hospital

Homer Pien *
Manager, Biomedical Technologies, and Head, Image Recognition Systems Laboratory
C.S. Draper Laboratory

Peer M. Portner
President, Novacor Division
Baxter CVG

Richard A. Rettig
Senior Social Scientist
RAND

Edward B. Roberts
School of Management
Massachusetts Institute of Technology

Nathan Rosenberg
Center for Economic Policy Research
Stanford University

Stephen I. Shapiro
Managing Director
The Wilkerson Group, Inc.

Kenneth Shine
President (until 2002)
Institute of Medicine

John S. Taylor
Director of Research
National Venture Capital Association

Frederick Telling *
Vice President
Pfizer

Samuel O. Thier
President and CEO
Partners Health Care Systems, Inc.

John T. Watson
Acting Deputy Director
National Heart, Lung, and Blood Institute

NAE Program Office Staff

Tom Weimer, Director
Proctor Reid, Associate Director
Nathan Kahl, Project Assistant
Robert Morgan, NAE Fellow and Senior Analyst

*Panel member

4

Report of the Panel on the Aerospace Industry

This report was prepared by the Panel on the Aerospace Industry, one of five panels formed by the Committee on the Impact of Academic Research on Industrial Performance. The panel of five included three members of the NAE (one from academia and two from industry), one other member from academia, and one from industry. Two of the NAE members were also members of the parent committee. The charge to the panel was to evaluate the past impact of academic research on the performance of the aerospace industry and identify ways to increase the impact in light of recent and ongoing changes in the structure and economic situation of the industry. The report is intended for policy makers in industry, government, and academia. Industry performance was defined as shareholder value. This metric differs from the traditional measure of success in the aerospace industry, which was its contribution to national security or to the space program.

The aerospace industry was selected for study as an example of an industry, now relatively mature, that developed with extensive funding by government in research and technology and that is dependent on advanced technologies for its present and future economic competitiveness. Therefore, the industry might provide a baseline for comparison with other less mature industry sectors.

The aerospace industry has been the beneficiary of more than 50 years of government-subsidized research conducted by industry, universities, and government laboratories. Subsidies have taken the form of direct funding by the National Aeronautics and Space Administration (NASA) and the U.S. Air Force, other defense-related funding, incentives in government contracts, and tax incentives; research was focused primarily on improving performance to meet the needs of national defense and the space race. In recent years, large cuts in federal support, combined with other competitive and financial pressures, have resulted in major changes in the industry.

Dramatic consolidation, largely the result of huge cuts in defense spending and greater emphasis on the use of commercial, off-the-shelf technologies in the 1990s, as well as increasing global competition, has changed the scope, priorities, and practice of aerospace R&D. Spending on R&D in the industry declined throughout the 1990s to less than half its peak in 1987. In 2000, total R&D spending (by government, industry, and other institutions) in aerospace totaled about $10.3 billion, accounting for roughly 9 percent of R&D among manufacturing industries (NSF, 2001a). Employment is down 40 percent from its peak in 1989, and the number of scientists and engineers in 1999 was less than half the number employed in 1986 (AIA, 2001). (It is interesting to note, however, that over the last two decades scientists and engineers in the aerospace industry have earned more than 25 percent more than their counterparts in other industrial sectors [AIA, 2001].) As technologies have matured, margins have shrunk, cost reductions have taken precedence over improvements in performance, and electronics and information technology now account for a large percentage of aerospace product value and technical emphasis. Priorities in R&D have changed accordingly.

Historic patterns of industry-university interaction, which were based on significant government funding of R&D, have been broken; new models will certainly emerge that encompass not only R&D funding, practice, and expectations, but also engineering education. But first, significant cultural and practical barriers will have to be overcome. Indeed, for academic research to have the maximum beneficial effect on the new aerospace industry, the entire structure of academic research in aerospace will have to change.

SCOPE OF THE STUDY

According to government classifications, the aerospace industry includes aircraft (NAICS 336411), aircraft engines and engine parts (NAICS 336412), aircraft equipment and parts (NAICS 336413), missiles and space vehicles and parts (NAICS 336414), guided missile and space vehicle propulsion units and parts (NAICS 336415), and guided missile and space vehicle parts not classified elsewhere (NAICS 336419). The panel has expanded this definition to include space-based information systems, a burgeoning segment of space commerce. The academic community that supports the industry was also defined broadly to include departments of mechanical engineering, materials science and engineering, and computer science, as well as the relatively few departments of aerospace engineering.[1]

Overall industry sales for 2000 exceeded $146 billion (U.S. Bureau of the Census, 2002). The panel considered it essential to limit discussion to five sectors of the industry that made significantly different contributions to total industry shipments of 1999–2000.[2]

- gas turbine propulsion systems: $10 billion
- civil transport aircraft: $30 billion
- launch vehicles: $11 billion
- unmanned aerial vehicles: less than $1 billion
- space-based information systems: $12 billion

Because the first three are mature sectors, cost and reliability have replaced performance as their principle criteria for success. The last two are relatively immature sectors that are undergoing rapid development and are, therefore, more dependent on new technologies.

Gas-Turbine Propulsion Systems

The gas-turbine industry developed in the 1950s and 1960s with the rapid conversion of both military and commercial aircraft from reciprocating to jet propulsion engines. Initially, several companies entered the arena, but in a rather short time all but about a half-dozen had dropped out, either because of the large financial investment required or because of technical difficulties. Currently, three large manufacturers of jet engines—General Electric Aircraft Engines, Pratt & Whitney, and Rolls-Royce—are engaged in intense competition for both military and commercial business. Smaller firms supply niche markets, such as general aviation or missiles.

Driven by competition for higher thrust/weight ratios and lower fuel consumption, the providers of jet engines are pressing materials and fluid mechanical and solid mechanical design procedures to their limits. In this high-stakes business, the development of a new engine costs up to $1 billion, and companies are eager to take advantage of the improved understanding and new techniques that academia can provide. The knowledge base for academic researchers is arcane, however, and the community of researchers is small. Presently, only about a half-dozen universities in the United States are making significant contributions.

Civil Transport Aircraft

Prior to World War II, the commercial aircraft industry was robust. With the introduction of the jet transport in the 1950s, the industry entered a period of rapid growth. Since the late 1970s, sales of civil aircraft have more than doubled, from $14.3 billion to more than $38 billion (AIA, 2001). Like the engine industry, the commercial aircraft industry initially comprised several companies, all but one of which have now been absorbed into the Boeing Company, which has dominated the field for the last decade. In fact, because some aerospace firms that are not formally part of Boeing manufacture major components of Boeing aircraft, its dominance is actually greater than it appears. Boeing's only substantial competition in commercial aircraft is Airbus Industry, a European consortium.

The commercial transport market is fiercely contested, and market share can be gained or lost by small differences in performance or cost. Technologies to improve performance and manufacturing are both critical to success, which creates a fertile field for academic research. A significant contribution of academic research to the industry has been the development of design techniques based in computational fluid dynamics. For example, the Aerospace Design Program at the Georgia Institute of Technology receives financial support from a number of aerospace companies. In addition, the Lean Aerospace Initiative at the Massachusetts Institute of Technology (see Box 4-1) and the Automation and Robotics Research Institute at the University of Texas-Arlington are addressing manufacturing technology and management issues and have attracted significant industry interest.

BOX 4-1
Lean Aerospace Initiative at the Massachusetts Institute of Technology

The Lean Aerospace Initiative (LAI) is a research program of the Center for Technology, Policy, and Industrial Development and the Department of Aeronautics and Astronautics at the Massachusetts Institute of Technology (MIT). LAI was launched in 1993 by leaders of the U.S. Air Force, MIT, defense aerospace businesses, and labor unions who recognized the potential of applying lean manufacturing and management concepts to the aerospace industry. Lean manufacturing practices, developed primarily by Toyota and first documented by James Womack, Daniel Roos, and Daniel Jones in *The Machine that Changed the World* (1991), have led to significant improvements in the automobile industry in terms of cost, quality, and productivity. The elements of lean practices being addressed by LAI include integrated product and process development, optimized product flow, total quality management in an environment of continuous improvement, and strong employee participation.

Participants in LAI are members of a consortium of government, industry, labor, and other universities. Members include the Aeronautical Systems Center and Space and Missile Commands of the U.S. Air Force; the Office of Safety and Mission Assurance of the National Aeronautics and Space Administration (NASA); The Boeing Company, General Electric Aircraft Engines, Hewlett Packard, Lockheed Martin, Northrop Grumman, Pratt & Whitney, Raytheon, Rockwell, Rolls-Royce Allison, and TRW; the International Union of United Automobile, Aerospace and Agricultural Implement Workers of America; the Aerospace Industries Association; and the University of Chicago, University of Washington, and Wharton School of Business at the University of Pennsylvania.

LAI represents a new model for research at MIT. The research agenda and priorities are developed jointly by MIT researchers and customers, and users of

Launch Vehicles

The launch-vehicle industry grew out of the ballistic missile industry with the dawn of the space age in 1958. The new industry was entirely dependent on government funding until the birth of the communications satellite industry in the 1970s. In the last 10 years, the commercial market for satellites has become comparable to the government market for satellites, missiles, and NASA mission hardware, and with the proliferation of communications satellite constellations, the demand for launch services has grown rapidly.[3] There are currently three major U.S. suppliers of launch services (Boeing, Lockheed Martin, and Orbital) and four foreign suppliers of either launch vehicles or launch services.

All but the two or three most recently developed launch vehicles have been derived from ballistic missile technology, which was heavily funded by the federal government from the 1950s through the 1970s. Initially, academic research made substantial contributions in key technical areas, such as reentry,

the research participate directly in the research process (e.g., case studies, benchmarking, and surveys). There is a great deal of interaction between sponsors and researchers, and research objectives are established by consortium members; specific research plans and approaches are developed by MIT researchers and reviewed and prioritized by focus teams of members. An LAI Research Council reviews and coordinates the research plans to ensure that the total research program meets stakeholder expectations. Research plans have been established for factory operations, product development, policy and external environment, supplier relations, and test and space operations.

Research results are organized into a systematic framework called the lean enterprise model (LEM), which is used to communicate results to consortium members and is a model and catalyst for change in the defense aircraft industry. LEM encompasses lean enterprise principles and practices derived from surveys, case studies, and other research activities, and provides a framework for disseminating LAI research results, a reference tool for consortium members to benchmark their lean attributes, and a tool to encourage the development of new lean paradigms in the design, development, and production of military aerospace products. For consortium members, LEM provides a guide to translating conceptual principles into practical applications.

LAI has developed a unique approach to academic research in aerospace because it is not focused on a particular technology or product but addresses cross-cutting topics throughout the aerospace enterprise, from customer and initial design to supply chain management and final product delivery. Research and the dissemination of results depend on the active participation of consortium members. Broad consortium membership from government, industry, labor, and other universities is also unique. The program has been of great value to sponsors and may provide a model for other research initiatives in the aerospace industry to address shortcomings identified by industry.

the combustion of rocket engines, and the development of resin matrix composites, which are widely used in solid-rocket motor cases, nozzles, and core vehicle components.[4] In the past few years, launch vehicles have not been a very fertile field for academic research, partly because of the lack of new developments and partly because the major problems have been developmental rather than basic.

Unmanned Aerial Vehicles

Although drones have been used since the beginning of aviation as targets and as research vehicles, unpiloted aircraft became important militarily with the introduction of cruise missiles enabled by the development of very small gas-turbine engines and terrain-following guidance systems, which extended their range and gave them the capability of attacking designated targets. Unmanned aerial vehicles (UAVs) have been operational for about two decades. Recently, however, there has been a good deal of interest in both the military and scientific communities in the development of very small, autonomous, aerial vehicles (microair vehicles [MAVs] so small that they are essentially covert) equipped with miniaturized imaging systems and guidance systems enabled by microcircuit technology (see Box 4-2).

BOX 4-2
Microair Vehicles

In 1994, Charles Ellington, a zoologist at Cambridge University, published the results of his research on insect flight, which revealed a microscale vortex at the leading edge of the wings during the downstroke. These findings were brought to the attention of the Defense Advanced Research Projects Agency (DARPA), which had begun a $35 million program to develop microair vehicles (MAVs). Dr. Ellington's insight into the aerodynamics of insect flight was the basis for DARPA's Mesoscale Machines for Military Applications Program, which began in 1998. The program provided $20 million over a period of three years to a handful of research institutions, including the Georgia Institute of Technology and Vanderbilt University, as part of a larger program to create versatile, robust reconnaissance "bugs." The research team at Georgia Tech, now joined by Dr. Ellington, and working with the Ohio Aerospace Institute, has developed and is seeking a patent for a "reciprocating chemical muscle"—a chemical power source that would power the wings, guidance system, and payload for a device called an entomopter. The team has achieved wing motion of 70 cycles per second with enough power to fly. Work continues on the Entomopter, with researchers also exploring its application for research and mapping of Mars.

Source: Toon, 2001.

Another area of growing interest is unpiloted vehicles capable of long endurance flight at very high altitudes for atmospheric sampling or surveillance. This type of UAV, marketed by innovative companies such as Aurora Flight Systems, was facilitated by advances in lightweight materials, control technology, and modeling and simulation. Several firms that emerged directly from academic research continue to rely on research by academics.

Space-Based Information Systems

Space-based information systems include all systems that use orbital assets to acquire or transmit information, such as observational satellites for military surveillance, weather satellites, navigation and positioning satellites, and commercial communications satellites. The first space-based information systems were military surveillance satellites that were launched in the 1960s with great secrecy. These systems have been systematically upgraded since then and are still an essential component of our national defense. Weather satellites and early geosynchronous communications satellites came next. The number and capacity of geosynchronous satellites have increased steadily. The largest potential increase in communication satellites, however, will be the launch of large constellations of satellites, some in low-Earth orbit ("little" and "big" LEOs) and some in specialized orbits for particular markets. Although several companies addressing this market failed in the late 1990s (e.g., Iridium, ICO Global Communications), others (e.g., Teledesic and Satellite LLC) plan to launch extensive satellite networks to provide broadband data communications.[5]

The increase in the number of satellites has been enabled by the explosive development of microcircuits, which made information processing possible, including information buffering and the passing of information between satellites in an array and ground stations. Electronic miniaturization has also made it possible to build satellites with considerable capabilities that weigh as little as 90 pounds. The key technologies for these satellites are microcircuitry, antennae, photovoltaic power supplies, lightweight structures, and small propulsion systems for orbital positioning and maintenance. High-volume, low-cost manufacturing of standardized parts has been essential to the emergence of this segment of the industry.

INNOVATION

The most obvious innovations in aerospace have come from industry through the development of new products and systems. This pattern goes back to the Wright brothers, who were motivated to conduct their research on airfoils and control by a desire to build and market a useful aircraft. With few exceptions (e.g., Robert H. Goddard, who was a professor, and C.S. Draper, who invented and developed inertial guidance), key innovations have not arisen from academic research. The pioneers of aerospace were engineers and entrepreneurs who used available technologies to create new capabilities for flight in the atmosphere or beyond.

Several government laboratories were created to stimulate the growth of the industry. The first of these was the National Advisory Committee for Aeronautics (NACA), which was created in 1915 when decision makers in the government became aware that the United States was far behind European nations in the development of aeronautics. The establishment of the Langley Aeronautical Laboratory at Hampton, Virginia, followed; research there was initially focused on aerodynamics, structures, control, and propulsion for military aircraft. The Langley laboratory had very little connection with universities. During World War II, some universities established very large and effective R&D programs. Examples include the Radiation Laboratory at MIT, which played a major role in the development of radar in the United States, and the nuclear laboratories of the University of California. It is important to realize that these were essentially industry laboratories embedded in the academic environment for the duration of the war. They did not pursue academic research agendas as we think of them now.

After World War II, all of the military services established laboratories to work on the technologies most important to them. These laboratories were staffed by a mix of civil servants and military personnel. One of the functions of the military-service laboratories was to maintain contact with universities by providing financial support and encouraging faculty to address issues of concern to the services. At about the same time, the services established organizations devoted to funding basic research. The first of these, the Office of Naval Research, was followed by the Air Force Office of Scientific Research (AFOSR) and the Office of Army Research. In their heyday, these offices commanded sizeable research budgets and funded much of the academic research in aerospace, a good deal of it only tenuously connected to the needs of the services.

In 1958, NASA was established with the U.S. commitment to the Apollo Program; NASA followed a similar pattern and funded a great deal of academic research. Whereas NACA had been almost entirely an in-house research organization devoted to aeronautical technology and facilities, NASA issued grants and contracts to industry and universities, using its civil service workforce to manage these activities. Of course, these research activities were subordinate to NASA's main mission, which was to go to the Moon.

Today, government agencies command far fewer resources for research and have focused more tightly on their missions. Some are also reexamining their relationships with academic researchers. For instance, in 1994 NASA's Office of Aeronautics appointed a University Strategy Task Force to review the agency's support of academic research and recommend policy and other changes to ensure the long-term health of aerospace research in academia.

Because of the combination of substantial government funding and the emergence of government laboratories dedicated to aerospace research, universities assumed a role of supporting the R&D activities of the services and NASA by addressing issues that emerged during the technology development process and improving the base technologies for later applications. Universities also played a large role in the development and improvement of techniques for analyzing fluid flows and structures that were enabled by advances in digital computation. But even in the area of computational fluid dynamics, much of the innovation originated in government laboratories and industry, with academic research playing a supporting role.

FINDINGS

Responses to questionnaires and discussions at a workshop convened as part of this study revealed consistent concerns about the impact of academia on the aerospace industry and the future of the university-industry relationship (see Addendum). These concerns can be divided into five subject areas: (1) the implications of changes in the federal research support structure; (2) the value placed on academic research and education by industry and the implications for industry support of academic research; (3) the impact of changes in the industry on the research and educational capabilities of universities; (4) intellectual property rights and how they affect university-industry collaborations; and (5) arrangements to promote cooperative research between academia and industry.

Changes in the Federal Research Support Structure

For three decades after World War II, the mission-oriented agencies of the federal government had charters from Congress and successive administrations to support a broad range of research without having to demonstrate the applicability of the research to their missions. The first agency with this kind of flexibility was the Office of Naval Research, but eventually the Air Force, Army, NASA, U.S. Department of Energy, and other smaller agencies were given the same support and flexibility. Later, the National Science Foundation (NSF) and the National Institutes of Health (NIH) were created with specific responsibilities for funding so-called basic research (meaning research without specific, known applications). NSF has not played a very significant role in funding for aerospace-oriented research, although some research on manufacturing funded through the NSF Engineering Directorate may be applicable to aerospace.

The strong support of the federal government for aerospace research has diminished significantly in the last decade. Federal funding has fallen dramatically, and available funding is being managed much more carefully to achieve specific results at lower cost. Mission-oriented agencies now insist on demonstrated relevance to their missions as a precondition for funding research. This represents a significant change from the earlier mind-set when ensuring the health of academic research was considered an important part of an agency's mission.

Beginning in 1990, the AFOSR began to assess academic research in terms of improvements in the performance of military aircraft. Universities were no longer considered AFOSR customers, but means to an end. AFOSR focused particularly on how the results of academic research could be transitioned into applications that would improve Air Force systems. For example, in fluid dynamics, a mature technology, the transition has taken 20 or more years from basic research to application. AFOSR is working toward shortening the transition time using a new model of technological innovation based on a work published by Stanford University, *Conceptual Foundations of Multi-Disciplinary Thinking* (Kline, 1995). Today, when the Air Force needs a new product, the reserve of knowledge is first reviewed, and research is funded only if the necessary knowledge is not available. AFOSR has found that the knowledge base created by years of support for research is extensive and that the need for new basic research is not as great as it once was. This reflects the maturity of the technology/product/industry. Using this approach, AFOSR program managers have become brokers between industry and academic researchers, facilitating networking in the research community by communicating the needs of industry, the sources of knowledge from past research results, and the creation of new knowledge through new research, when necessary.

These changes have been made in other agencies as well. NASA, for example, used to have a generic space technology research program. Today, research decisions have been distributed to offices with mission responsibilities, and only technologies with near-term mission applicability are supported.

Because of these changes, significantly less advanced research in aerospace is being done or even contemplated by the U.S. government, potentially shrinking the future pool of technology available in the field. Arguably, this situation may be a correct response to the maturing process; technological progress in the industry is progressing slower than in the past; the development of new aerospace systems is now measured in decades rather than years.

Value Placed on Academic Research and Education

Industry's Viewpoint

The panel's research strongly suggests that mature sectors of the industry value academia principally for its graduates at all levels. Therefore, industry places more value on researchers than on research itself. Masters-level students are especially attractive to industry because they have a broad knowledge of their fields and have not become specialized in a narrow area of interest required for a doctoral thesis. Although numerous technical contributions from academia can be identified (see Box 4-3), the research results per se are not highly valued, partly because improvements in industry performance (defined as increasing shareholder value) can seldom be traced to them. The indirect benefits of academic research are often not recognized by industry, despite their contributions to the knowledge base that may ultimately contribute to the development of new technologies.

The contribution of academic research to the development of very large-scale integration is one example. Academic research directly benefited chip makers (e.g., Intel) and indirectly benefited aerospace companies by enabling the development of enhanced avionics. However, research results become visible to management only when they are applied directly to an aerospace system or improve a company's performance. Another example is the development of composite materials, the underpinnings for which were developed in academia and the applications of which include turbine engines, solid rocket motors, and other critical and noncritical engine and airframe components. The indirect benefits may not be recognized, although they have tremendous value. In contrast, university graduates are highly valued because they provide direct visible benefits.

If academic research is considered in the broad context of innovation processes, it plays a very large role, along with government laboratories, in laying the foundations of understanding that lead to the next wave of innovation. This role is reflected in recent studies by Diana Hicks and Francis Narin (2001) of CHI Research and others (e.g., Spencer, 2001) showing the predominance of academic research papers cited in patent applications. Of the scientific papers cited on U.S. industrial patents in 1993–1994, academic research was the source of 52.1 percent, roughly twice the number of industry research papers (Narin et al., 1997).

In the current environment, future opportunities for academic research will be much more focused on issues that contribute to market success and will require much more flexibility on the part of academic researchers. For industry, the focus of R&D will be success in the marketplace; R&D should generate technological discriminators, improve affordability, and create new business opportunities. To receive industry support, academic researchers will have to meet industry's needs for timely and usable results. To achieve these results, industry will have to manage its relationships with universities more effectively.

Typically, large aerospace companies define key corporate strategic opportunities and fund them first. Subsequent R&D spending authority is dependent on profit and loss. With a few exceptions, large aerospace companies no longer have central R&D laboratories. Instead, they use contract research capabilities when they are available, and they fund universities philanthropically to support centers and institutes for research in specific areas. At this stage, aerospace companies are especially keen on the development of better methodologies and tools, such as tools in computational fluid dynamics and the research results of the Lean Aerospace Initiative.

Small companies maintain a very different relationship with universities for a number of reasons. First, there is a significant state and federal

BOX 4-3
Examples of Academic Contributions to Aerospace

Although the consensus view of the respondents to the panel's questionnaire is that academic research has contributed mostly indirectly to the performance of the U.S. aerospace industry, a wide array of technologies were identified to which academic researchers had contributed in meaningful ways. In some cases, the contributions were the development of tools, especially models of system behavior; in other cases, specific technologies were cited.

Tools
- radar cross-section modeling
- extrinsic Fabry-Perot interferometer (EFPI) applied to smart structures
- models of satellite communication systems
- performance of competing aircraft design tools
- modeling of electromechanical theory
- advanced nonintrusive instrumentation
- flow-visualization techniques
- computational fluid dynamics
- microelectrommechanical systems(MEMS)
- multidisciplinary optimization
- nonlinear, constrained optimization codes for designing control laws
- modeling of the mechanical properties of composites
- thermomechanical fatigue testing
- electron backscattering technique for fine-scale analysis of crystal orientation
- thermodynamic software for predicting phase equilibria (Thermocalc)

infrastructure to support small business research, especially the Small Business Innovation Research (SBIR) Program and state programs that provide loans and matching grants. Second, many states and universities have technical business "incubators" that provide technical and management expertise for new technology companies until their ideas are commercially viable. Once these firms graduate from the incubators, they have multiple connections to the university for continued research. Third, small firms typically have little or no research staff; contracts with universities, faculty consultants, student cooperatives, and other resources provide a cost-effective mechanism for securing technical expertise. Fourth, maintaining close relations with universities gives small businesses access to information and cutting-edge technology that they often have little time to obtain on their own.

Applied Research
- heat transfer, combustor cooling, aeromechanics
- vane-blade interaction for transonic turbines to understand unsteady flow
- influence of clouds of volcanic ash on turbine engine performance
- low Reynolds number airfoil design
- Internet by satellite, including protocols and computational tools for data integration
- folding-wing design for small UAVs
- fracture and flaw sizes of critical brittle materials
- aluminum-lithium alloy applications in space systems
- amorphous alloys
- obstacle detection and avoidance, relative navigation, mapping
- differential absorption lidar

Basic Research
- theoretical basis for UAV flight controls
- Shannon's information theory
- electromagnetic antenna theory
- linearized unsteady flow analysis
- composite laminate theory
- improved understanding of fiber-matrix interactions in composite materials
- superplasticity
- real-time decision systems using artificial intelligence

Centers
- Gas Turbine Laboratory (MIT)
- Turbomachinery Components Research Laboratory (Iowa State)
- Whittle Laboratory (Cambridge University)
- Lean Aircraft Initiative (MIT)
- Automation and Robotics Research Institute (University of Texas-Arlington)

Finally, university research provides intellectual mixing that leads to constant discussions and debates, both formal and informal, about new ideas. Universities serve as brain trusts that provide a long-term perspective on ideas and technology development that is difficult to find elsewhere and is exceedingly valuable to small firms. Small manufacturers of UAVs illustrate many of these points (see Box 4-4).

However, industry must learn how to manage relationships with universities. Small companies tend to want technical expertise from universities, while larger firms tend to want system engineers (with technical expertise). Because of academic schedules, universities are not well suited for research on the short-term critical path. For longer term projects, however, academic research can provide the foundation for new products and services that may be critical to a company's business.

BOX 4-4
Unmanned Aerial Vehicles and Academic Research

The unmanned aerial vehicle (UAV) sector is representative of the high-risk, still-emerging end of the aerospace industry. A brief description of three companies developing new aircraft in this sector illustrates the range of experiences companies can have with universities. These examples also demonstrate the dangers of making broad generalizations about relations between industry and universities in the aerospace industry.

The Cypher UAV, developed by Sikorsky to meet U.S. military and commercial needs in ground and naval surveillance, relaying communications, forestry surveillance, law enforcement, and search-and-rescue missions, recently became operational with the U.S. Marine Corps. Cypher combines the efficiency of a ducted airstream with the attributes of a coaxial, advancing-blade rotor system. It can take off and land vertically, carry a payload of about 50 pounds for several hours, and cruise at 80 knots. Academic involvement in the development of Cypher has been intermittent (e.g., in the development of the panoramic view sensor). Because of the small amount of government funding available to support R&D on rotary aircraft, work with universities has proven to be difficult, and most contacts with academics has been initiated by universities rather than program managers. Managers pursued ideas to see if a common interest could be identified, but often academic interest was primarily a search for funding to support research and graduate students. Most companies believe they derive little benefit from working with universities unless a common interest can be found because of problems in meeting deadlines, the frequent turnover of students working on projects, and a tendency for academics to promise more than they can deliver.

Aurora Flight Systems was formed in the 1980s by an MIT graduate and a Harvard graduate to build robotic aircraft to study chemicals in the upper atmosphere. According to company president John Langford, Aurora depends on universities for its research and considers funding of academic research a way of buying a time-share in the world's best research laboratories. Aurora has

The reexamination of university-industry relationships should not be confined to research. Engineering education must also be adapted to an environment in which the context of engineering, the workplace, and the role of technology have changed dramatically. For instance, the ethics of engineering are becoming more complicated. Mission success may conflict with the emphasis on market success, which demands sticking to business-determined costs and schedules. The required capabilities of engineers, especially their ability to work in multidisciplinary teams, are also changing. Many have argued that engineering colleges have not responded well to these changes (see Box 4-5).

Academic Viewpoint

The principal measure of success for a research university is the quality of its faculty and student body. Faculty success is measured principally by standing in partnerships with universities on Small Business Innovation Research (SBIR) projects and has taken advantage of state programs to encourage cooperation with state universities. The company considers academic research a way of laying the foundations for new products and services, rather than a mechanism for solving short-term critical problems.

Aurora has contracts with universities in many key technical areas: computational fluid dynamics and analysis codes; advanced propulsion concepts; differential absorption LIDAR; advanced navigation, guidance, and control algorithms; and composite material testing. The company works with MIT, West Virginia University, University of Vermont, University of Virginia, University of Maryland, and Virginia Polytechnic Institute. Aurora worked with MIT, for example, to develop fault-tolerant controls to eliminate redundant flight control system hardware, which adds weight to the aircraft. Because fault-tolerant control systems have applications in many industries, Aurora created a new business, Athena Technologies, to commercialize these technologies.

Freewing Aerial Robotics Corporation was the direct result of the technology incubator program at the University of Maryland. As a start-up in the incubator, Freewing had access to technical and business advice from the university, office space, university facilities, especially the wind tunnel and computer equipment, and work-study students. In return, the University of Maryland held a 3 percent share in the company. Freewing's first product was the Scorpion 100-50, a tactical, low-altitude, subsonic UAV designed from the beginning for military and civilian use. New products were planned based on a patented "tilt-body" technology that created a new class of aircraft applicable to any size and speed. However, inability to raise fresh capital forced Freewing to file for Chapter 7 bankruptcy in October 2001.

The University of Maryland and Texas A&M University contributed resources and research capabilities to the development of two critical technologies: (1) a nonlinear mathematical model of the aerodynamics of a freewing aircraft and subsequent wind tunnel testing; and (2) avionics for an autopilot.

research peer groups (in exceptional cases very innovative teaching can convey star status). The most prestigious faculty members seem to attract the brightest and most ambitious students. Thus, universities compete vigorously for the research elite who are the recognized intellectual leaders in their fields and are willing to expend significant discretionary resources to attract them for laboratories, graduate student support, and faculty control of research funds. Thus, major research universities do not depend entirely on external funding to support the research programs that determine their standing in the academic hierarchy. One measure of independence is that most leading research universities now pay their faculty almost entirely from internal funds, rather than through research contracts, even though faculty typically spend about half of their time working on funded research projects. Other costs of research include matching-fund requirements, costs for bids and proposals, and other costs that are not fully reimbursed

BOX 4-5
Strengthening Academic Support

For more than a decade, government agencies, the aerospace industry, and universities have expressed concerns about how well universities can meet the engineering needs of industry and, conversely, the willingness and ability of government and industry to provide the funding necessary for universities to maintain and enhance their research capability. Studies by the National Research Council (1995), the American Society for Engineering Education (1994), and others have explored the changing needs of industry and the ability of universities to meet those needs. Based on these studies, government has undertaken a number of initiatives, such as a project in the mid-1990s by the National Science Foundation to form coalitions of universities to address systemic reforms in the engineering curriculum and related issues. These activities reflect an awareness by all parties—industry, universities, and government—that the needs of industry are not being met.

In the aerospace industry, the need to improve relations with universities and to improve the quality of graduates led to at least two major initiatives in the early to mid-1990s, one by industry and one by government. Based on attempts in the late 1980s to improve relations with universities historically important to Boeing, the Boeing Company initiated what has become the Industry University Government Roundtable for Enhancing Engineering Education (IUGREEE) (McMasters, et al.,1999). IUGREEE held its first meeting, with additional industry and university participants, in 1995. The goal of IUGREEE was to catalyze actions, rather than echo recommendations of other studies.

IUGREEE now perceives itself as providing a broad-based industry voice for changes in engineering education. IUGREEE has adopted the following objectives:

- to articulate and draw attention to critical issues that industry believes should affect engineering education

by contracts. In the final analysis, universities contribute substantially to support for research rather than making money on research, as is often assumed.

Some of the brightest and most innovative young faculty may find that furthering their careers in academia often conflicts with focusing on developing strong cooperative relationships with industry. Indeed, too much emphasis on developing industry ties can seriously retard their progress. These same young faculty are often poorly informed about the needs of industry and the differences between industry and academic work modes. Thus, they do not provide role models for students, most of whom seek employment in industry after graduation. To address this problem, Boeing introduced an academic fellows program that brings 10 to 15 academics to work in the company during the summer. The goal of the program is to provide firsthand knowledge of how industry operates and to expose academics to real-world technical problems. Leaders at Boeing argue that the only way to overcome long-standing cultural differences between

- to develop agendas to see that changes are made
- to facilitate the implementation of these agendas through existing organizations and other mechanisms

In 1995, in response to similar concerns, NASA's Aeronautics Advisory Committee appointed a Task Force on University Strategy to examine issues related to the role of universities in NASA's Aeronautics Enterprise. The task force was asked to address criticisms of NASA's management of industry and university relations and to recommend ways to improve collaborations between NASA's aeronautics program and universities. The Task Force on University Strategy undertook the following tasks:

- a review of NASA's Aeronautics Enterprise Mission Statement, Strategic Plan, and related documentation vis-a-vis U.S. universities
- a review of NASA's policies and practices toward research grants for the aeronautics program
- a comparison of university research strategies used by other government agencies and NASA's industry collaborators with the policies and practices of NASA's Aeronautics Enterprise
- a review of the data collected by previous informal study groups and symposia
- the development of recommendations to improve how NASA's Aeronautics Program collaborates with universities, including a better definition of the role of university research in the Aeronautics Enterprise and methods of supporting, evaluating, and collaborating in that role

In 1997, the task force issued recommendations on NASA's funding of basic research, the use of peer reviews, the protection of intellectual property, an increase in funding limits for multiyear grants, collaborative research with universities and industry, and other aspects of relations with universities (NASA, 1997).

industry and academia is to provide opportunities for interaction and to maintain a dialogue to identify differences and develop potential solutions.

Much of the tension between the needs of the academic research system and industry's needs for personnel and technology can be addressed by better management of university-industry relationships. For example, academic involvement in problem definition can help focus research on meeting the needs of both. Industry must also recognize that universities cannot work to an industry schedule and should choose research projects accordingly. Projects must be interesting enough to attract graduate students and must be structured in a way that is compatible with the university reward system. A combination of long-term and short-term funding can provide stability for academic programs that may then be in a better position to address shorter term industry problems. Active industry advisory boards can help overcome communication problems.

Academia also has constraints that industry does not always comprehend. Universities have costs and business pressures just like corporations, although in a different context. Because of reductions in government funding and more frequent matching-fund requirements, the dollar amounts of individual government research grants have been reduced. Industry grants also tend to be small, averaging about $43,000 for one large aerospace firm. Chasing many small grants can be expensive. Wealthier research universities can absorb these costs more readily than most, but the costs are real. Universities may also face legal restrictions imposed by many states on fee-for-service research contracts that might put the university in direct competition with a private research firm. To industry, these types of constraints can be frustrating and may make universities seem unresponsive. Industry must recognize these constraints and modify its expectations.

Foreign Students

A large fraction of graduate students from around the world who attend U.S. universities remain in the United States, attracted to the innovative, dynamic, research environment in this industry. In 2000, of the 215 doctorates awarded nationwide in aeronautical and aerospace engineering, 85 were awarded to non-U.S. citizens, only 10 of whom had permanent visas (NSF, 2001b). Export laws and other restrictions, however, limit the involvement of foreign students in some types of aerospace research. Typically, for companies that have a substantial fraction of business with the DOD or NASA, employing noncitizens is difficult. Even for purely commercial applications, companies are wary of funding academic research performed by foreign students who may take the results to foreign competitors after they graduate. In some cases, firms are legally required to restrict the involvement of foreign students on contracts with universities. Generally, universities find it difficult to accept these restrictions because they cannot guarantee the future use of information generated in their laboratories. Some universities have restricted some of their laboratories to graduate students who are U.S. citizens, but these arrangements may become increasingly difficult to maintain as the percentage of foreign students (and faculty) in science and engineering school populations increases. Therefore, industry and government may have to consider changing their policies to ease restrictions on foreign students and faculty, many of whom opt to remain in this country and thus contribute to U.S. competitiveness (NSF, 2001b).

Intellectual Property Rights

Another key barrier to collaboration between universities and industry is intellectual property rights. Within universities, intellectual property is usually shared by the faculty member and the university; although the professor may be willing to negotiate ownership away, the institution tends to be very reluctant to do so. For obvious commercial reasons, companies generally insist on proprietary rights to research results produced by programs they fund. Universities, for a variety of reasons, find it either not in their interest or impractical to grant such rights and, perhaps more importantly, to protect them in the courts. As a result, companies must cope with different policies at different universities and even different projects at the same university. Pursuing the negotiating process on a case-by-case basis can be costly.

University faculty and administrators sometimes have inflated expectations of the value of intellectual property, which must be considered in the context of how its value can be realized. If the university and/or faculty member retains ownership but cannot sell it or develop a product from it, the value is much lower than if ownership is assigned to the company, and the proceeds are shared in some negotiated way.

Ownership of intellectual property is not the only issue that must be addressed. Mechanisms are also needed to protect intellectual property, especially in universities with a culture of disbursing knowledge. In Georgia, state laws were changed to allow Georgia Tech to protect intellectual property for three years, which is usually long enough to meet commercial needs. This arrangement has the added advantage of teaching graduate students to respect intellectual property as they will be expected to do in industry.

Arrangements to Improve Research Collaborations

In the course of committee and workshop discussions, a number of suggestions were made for improving research collaborations between universities and industry. Representatives of large and small companies and academics had quite different ideas. Generally, large companies called for changes in the structure of academic research organizations to make them more compatible with industry needs; universities called for more flexibility on the part of companies and a recognition of the limitations under which they operate.

Research Partnerships Abroad

Some of the problems associated with industry-university collaboration might be solved by the formation of exclusive, one-on-one research partnerships; industry partners would conduct research, in return for support from the single, industrial partner, and universities would give up the right to accept support from other companies. An example of this approach is Rolls-Royce's university technology centers (UTCs), a program established in 1990 to focus and coordinate Rolls-Royce university research in the UK. The firm targeted university departments with proven track records and expertise in technical areas of strategic importance to Rolls-Royce. UTCs at 12 universities in the UK conduct about half of the research funded by Rolls-Royce across a wide spectrum of short-term and long-term needs. The UTCs have quarterly management meetings and annual reviews with Rolls-Royce managers, and they compete for funding to ensure that the company gets value for its money. UTCs are also part of the firm's recruitment and training strategy, ensuring access to the best people the UK with the necessary technological training. University Technology Partnerships (UTPs), intended to complement the UTCs, involve suppliers, customers, and other European universities. For example, the Rolls-Royce UTP in Engineering Design Processes involves British Aerospace and the Universities of Sheffield, Cambridge, and Southampton. It was launched in 1998 to conduct a joint program of research into engineering design processes for the twenty-first century.

U.S. Research Partnership

One of the presentations at the workshop described the efforts of GE Aircraft Engines (GEAE) to develop a new strategy for funding academic research, similar to the strategy used by Rolls-Royce but tailored to the issues specific to U.S. universities. The strategy, called the University Strategic Alliance (USA), is intended to shift the company's current pattern of funding of 140 small contracts at many universities to funding of much larger contracts at a few universities that could become long-term partners. The change was partly motivated by a 66 percent drop in the company's internal engineering staff, which has made the company anxious to make more effective use of university capabilities.

GEAE's idea behind USA is to integrate university research into the firm's business strategy and technology road maps. The company would enter into a guaranteed, performance-based, five-year contract with sufficient funding to ensure a critical mass of capability at each university. Contracts would be focused on specific problems, such as film cooling issues on high-pressure turbine airfoils, in real-world contexts. The company's intent is to ensure reasonable academic freedom, reasonable protection of intellectual property rights, and opportunities for publication (with some restrictions). The universities would have access to company technology. The company would share ownership of intellectual property with the university but would generally not pay royalties.

Because protecting intellectual property can be a difficult issue, terms would be negotiated with each university. In some cases, the company might own the intellectual property rights but allow publication; in other cases, it might share patents. The same university could have different contracts for the treatment of intellectual property for different projects. A person with technical knowledge on each side would participate in the negotiations so that decisions would not be made solely by legal departments.

The company's goal is to enlist academic research to solve problems, many of which require fairly basic research. The company would allow university researchers to identify ways the problems could be solved, as long as those approaches could actually be implemented and would result in solutions in one to three years. The company envisions that faculty and students will come to its facilities to learn about problems in a real-world context and will pursue solutions in both company and university facilities.

From the company's point of view, this kind of company-university interaction would have significant benefits, particularly compared to the relatively little value the company now receives from the academic research it funds. From the university perspective, however, it may be difficult to participate despite assurances of long-term funding. A particular concern is that the problems posed by the company might not be attractive research subjects to faculty members because they would probably not involve problems at the intellectual frontier of the field and, therefore, would not be supported by the academic reward system. The arrangement might also be perceived as too restrictive in terms of publication and the university's freedom to work with other companies. Finally, the company may not gain access to the best researchers, who can raise research funds from other sources without the restrictions imposed by the company and so may not perceive any benefit from participation.

This example highlights some of the difficulties in improving university-industry research collaborations. The proposed strategy addresses the company's desires to have universities perform productive research and attempts to accommodate the university's need for open publication, protection of intellectual property, and stable funding. Nevertheless, U.S. universities may not consider this plan in a positive light.

CONCLUSIONS

Conclusion 4-1. Academic research in aerospace is changing rapidly. The government policy change reducing government support for aerospace in general means support for research in aerospace will also be reduced. Henceforth, market forces will determine how things change, and industry and universities will have to adjust accordingly.

Conclusion 4-2. Although industry values academia for turning out educated graduates and, to a lesser extent, for its research results, industry is not willing to support generalized research programs. Industry will provide support for programs with clearly identifiable impacts on its performance.

Conclusion 4-3. In the more mature parts of the aerospace industry (e.g., airframe and engine manufacturing), better methodologies and tools are the most valued research products. These include tools in computational fluid dynamics and the results of the Lean Aerospace Initiative. In less mature industry sectors (e.g., UAVs, MAVs, and satellite-information systems), new concepts and physical understanding will be most beneficial.

Conclusion 4-4. Industry considers academic research contributions most valuable (i.e., most relevant and easiest to transfer) when they result from cooperative programs with industry.

Conclusion 4-5. Issues concerning proprietary rights pose substantial barriers to improving collaborations between academia and industry. Resolving these issues will require compromises between industry's desire to protect intellectual property and academia's desire to maintain a free and open intellectual community.

Conclusion 4-6. Formal research partnerships between industry and academia are evolving. For new approaches to succeed, however, the academic reward system will have to find a way to recognize the value of close collaboration with industry.

NOTES

[1]Many engineering colleges have merged aerospace engineering departments with other departments, typically departments of mechanical engineering.

[2]Because the five subsectors studied by the committee do not align with the industry subsector definitions used to report sales and shipments data, these numbers are rough estimates that draw upon data from the following sources: AIA, 2001; McGraw-Hill/DOC, 2000; and CRS, 2003.

[3]In 1986, commercial launches accounted for only 13 percent of commercial and government launches (excluding space shuttle launches). By 1996, commercial launches accounted for half of the total (McGraw-Hill/U.S. Department of Commerce, 1998).

[4]Researchers at the University of Pennsylvania developed the underpinnings of laminate theory in the 1960s. Since then, Michigan State University, Case Western University, Stanford University, University of Connecticut, and University of Wyoming, among others, have contributed to the understanding of fiber-matrix interactions and to the modeling of mechanical analysis of composite materials.

[5]Satellite LLC purchased the assets of Iridium in late 2000 and relaunched operations of the 73 LEO satellites previously launched by Iridium. An additional five satellites were launched in February 2002. See *Washington Post*, February 12, 2002, p. E4.

REFERENCES

AIA (Aerospace Industries Association). 2001. Aerospace Facts and Figures. Washington, D.C.: Aerospace Industries Association.

American Society for Engineering Education. 1994. Engineering Education for a Changing World. Washington, D.C.: American Society for Engineering Education.

CRS (Congressional Research Service). 2003. Unmanned Aerial Vehicles: Background and Issues for Congress. Washington, D.C.: Library of Congress. Available online at: *http://www.fas.org/irp/crs/RL31872.pdf.* [August 21, 2003]

Hicks, D., and F. Narin. 2001. Strategic Research Alliances and 360 Degree Bibliometric Indicators. Pp. 133–145 in Strategic Research Partnerships: Proceedings from an NSF Workshop. NSF 01-336. Arlington, Va.: National Science Foundation.

IUGREEE. 2001. Vision 2010. Available online at: *http://www1.eng.iastate.edu/iugreee/.* [June 24, 2003]

Kline, S.J. 1995. Conceptual Foundations of Multi-Disciplinary Thinking. Palo Alto, Calif.: Stanford University Press.

McGraw-Hill/U.S. Department of Commerce. 2000. U. S. Industry and Trade Outlook '00. New York: McGraw-Hill.

McMasters, J., B.J. White, and T. Okiishi. 1999. Industry-University-Government Roundtable for Enhancing Engineering Education. AIAA 99-0281. Reston, Va.: American Institute of Aeronautics and Astronautics.

Narin, F., K. Hamilton, and D. Olivastro. 1997. The increasing linkage between U.S. technology and public science. Research Policy *(26)*3: 317–330.

NASA (National Aeronautics and Space Administration). 1997. Report of the University Strategy Task Force. Washington, D.C.: Office of Aeronautics, NASA.

NRC (National Research Council). 1995. Engineering Education: Designing an Adaptive System. Washington, D.C.: National Academy Press.

NSF (National Science Foundation). 2001a. Research and Development in Industry: 2000. NSF 03-318. Table A-3. Online at: *http://www.nsf.gov/sbe/srs/nsf03318/start.htm.* [June 24, 2003]

NSF. 2001b. Science and Engineering Doctorate Awards: 2000, S.T. Hill, author. NSF 02-305. Arlington, Va.: Division of Science Resources Statistics, National Science Foundation.

Spencer, J. 2001. How relevant is university-based scientific research to private high-technology firms? A United States-Japan comparison. Academy of Management Journal 44(2): 432–440.

Toon, J. 2001. Flying on Mars. Research Horizons 19(1): 19–23. Available online at: *http://gtresearchnews.gatech.edu/reshor/rh-f01/mars.html.* [June 24, 2003]

U.S. Bureau of the Census. 2002. Statistical Abstract of the United States—2002. Washington, D.C.: U.S. Government Printing Office. Available online at: *http://www.census.gov/prod/www/statistical-abstract-02.html.* [June 24, 2003]

Womack, J.P., D.T. Jones, and D. Roos. 1991. The Machine That Changed the World. New York: Harper-Collins.

ADDENDUM

Questionnaire

The following questionnaire was sent to selected individuals in various segments of the aerospace industry, some of whom attended the December 1998 workshop. Included among the questionnaire respondents were senior executives at Aerospace Corporation, Boeing, Draper Laboratories, GE Aircraft Engines, Lincoln Laboratories, NASA Lewis Research Center, Northrop Grumman, Orbital Sciences Launch Systems, Orbital Space Systems, Rolls-Royce Allison, SAIC, TRW, Inc., and professors with expertise in aerodynamcs, heat transfer, high-speed instrumentation, and turbomachinery from Iowa State University, Ohio State University, and Carnegie Mellon University.

STUDY OF THE IMPACT OF ACADEMIC RESEARCH ON INDUSTRIAL PERFORMANCE

1. Please identify the targeted sector or sectors of the aerospace industry with which you are most familiar.

- ☐ Space-based information systems
- ☐ Launch vehicles
- ☐ Transport aircraft
- ☐ Unmanned aerial vehicles
- ☐ Gas turbine propulsion systems

2. Please describe any cases you are aware of in which the contributions and impact of academic research to sector(s) have been clearly evident. What were the circumstances that led to the favorable outcome? (If possible, please supply references to published information.) *Please use additional pages, as necessary.*

3. Overall, would you describe the impact of academic research on industrial performance in your sector(s) of the aerospace industry as (please put an X in one box):

- ☐ 1. very large
- ☐ 2. large
- ☐ 3. medium
- ☐ 4. small
- ☐ 5. very small/nonexistent

4. The Panel has identified a number of mechanisms (listed below) by which it believes academic research has an impact on performance in the aerospace

industry. Please rate the importance of each mechanism using the same 1–5 scale used above [where 1 is "very large" and 5 is "very small/nonexistent." Where possible, please write down under each item some specific contributions you are aware of that have been made by academic research to industry via the mechanism. *Please use additional pages, as necessary.*

( ) A. Research-related Education and Training (of graduate students)
- ( ) i. at the M.S. level
- ( ) ii. at the Ph.D. level

( ) B. Invention and Innovation
For example, please consider any specific patents owned by a university or specific innovations resulting from university research that have been of benefit to your business or industry.

( ) C. Consulting
This refers to faculty providing expert advice to the industry.

( ) D. Technology Filtering
This refers to the role of academic research in helping companies identify research/technological opportunities and research/technological "dead ends."

( ) E. Tools and Productivity
This includes the development of analytical, computational, and experimental tools and methods that are adopted by industry.

( ) F. Pontification, Professionalism, Foundations
This refers to the function of faculty in documenting and presenting information critical to aerospace; in conveying relevant information to others in the profession, investors, and legislators; and to service to the profession, for instance, in organizing and sustaining professional societies, industry associations, and other trade groups.

5. Are there systematic trends that will either increase or decrease the impact of academic research in the future? For example, the Panel perceives that industrial in-house research is being de-emphasized as a result of downward pressure on engineering staff sizes. Is this a trend you perceive also? If so, will this result in more or less opportunity for productive research partnerships between industry and academe? Are there other trends of importance? *Please use additional pages, as necessary.*

6. What changes are required, if any, in academic research if it is to be responsive to trends and emerging challenges in the industry? *Please use additional pages, as necessary.*

7. To what extent is academic research relevant to aerospace justified for the research results it produces versus the academic infrastructure (faculty, post-docs, other research personnel, facilities) it supports, which enables the other contributions outlined above in question #4? *Please use additional pages, as necessary.*

8. Do you see any downside to enhancing university-industry research collaboration? Things to be avoided? *Please use additional pages, as necessary.*

9. Other comments? *Please use additional pages, as necessary.*

Workshop Agenda

WORKSHOP ON ENHANCING ACADEMIC RESEARCH CONTRIBUTIONS TO THE AEROSPACE INDUSTRY

December 4, 1998
National Academies Building
2101 Constitution Avenue, N.W.
Washington, D.C.

9:00 a.m. **Introduction to Study and Summary of Status**
J. Kerrebrock

9:20 a.m. **Summary of Findings from Questionnaire**
T. Mahoney

9:30 a.m. **Session I: General Discussion**

How have changes in the aerospace industry, as well as changes in academia, especially engineering, affected university-industry research cooperation?

How might the future of research cooperation be different from the past?

What role does academic research play in the total research enterprise (industry, university, government) in this industry?

10:15 a.m. Break

10:30 a.m. **Session II: Discussion of Specific University-industry Interactions: Presentations by Industry Participants**

- criteria to determine what types of research universities do best
- managing the research interaction: definition of deliverables, monitoring progress
- long-term vs. short-term collaborations
- single firm vs. consortia-based collaborations
- stellar examples of success and failure

12:00 p.m. Working Lunch

12:30 p.m. **Session II Continued: Presentations by Academic Participants:**

- sources of research funding
- preferred research for industry and expectations of industry
- managing interaction with industry, both for specific research projects and overall
- long-term vs. short-term collaborations
- single firm vs. consortia-based collaborations
- stellar examples of success and failure

1:30 p.m. **Session III: Detailed Discussions** (in breakout groups, if desired)

2:30 p.m. **Presentation and Discussion of Findings**

1. Can best practices be identified? Do they correlate to research with significant impact? Can the impact from academic research be clearly differentiated from the impact of internal and government research?

2. How might university-industry research collaboration be managed better to meet the needs of both?

3. How essential is industry participation in academic research, as a funder, definer, and active participant, both from the industry viewpoint and to meet the academic mission of research and education?

4. Is it possible to maintain the desired educational role of academia, in the absence of traditional academic research programs? If not, then how do we motivate and justify academic research, especially at the Ph.D. level, in the absence of a strong need for its research output?

3:30 p.m. **Formulation of Findings, Conclusions, and Recommendations**

4:30 p.m. Adjourn

Workshop Attendees

Jack L. Kerrebrock, *chair**
Professor of Aeronautics and Astronautics
Massachusetts Institute of Technology

William F. Ballhaus, Jr.
Corporate Vice President
Science and Engineering
Lockheed Martin Corporation

John S. Baras
Director, Center for Satellite and Hybrid Communication Networks
University of Maryland

Jewel B. Barlow
Director, Glenn L. Martin Wind Tunnel
University of Maryland

Robert A. Delaney
Chief, Design Technology
Rolls-Royce Allison

Robert M. Ehrenreich
Senior Program Officer
National Materials Advisory Board
National Research Council

Antonio L. Elias*
Senior Vice President, Advanced Programs
Orbital Sciences Corporation

Kenneth C. Hall*
Associate Professor
Department of Mechanical Engineering and Materials Science
Duke University

David Heebner (retired)
SAIC

Robert J. Hermann*
Senior Partner
Global Technology Partners, LLC

Wei H. Kao
Director, MMTC Structural Materials Department
The Aerospace Corporation

Kent Kresa*
Chairman, President, and CEO
Northrop Grumman Corporation

John S. Langford
Chairman
Aurora Flight Sciences Corp.

James McMichael
Program Manager
DARPA/TTO

Art Roch
Director, Advanced Technology
Commercial Aircraft Division
Northrop Grumman Corporation

*Panel member

Arun K. Sehra
Deputy, Aeronautics Directorate
NASA Lewis Research Center

S.K. Varma
Aerospace Consultant
Bethesda, Maryland

David C. Wisler
Manager, University Programs and Aero Technology Labs
GE Aircraft Engines

Consultant to the Aerospace Industry Panel

Thomas C. Mahoney
Director, WV-MEP
West Virginia University

NAE Program Office Staff

Tom Weimer, Director
Proctor Reid, Associate Director
Robert Morgan, NAE Fellow and Senior Analyst
Penny Gibbs, Administrative Assistant

5

Report of the Panel on the Transportation, Distribution, and Logistics Services Industry

The Panel on Transportation, Distribution, and Logistics Services Industry was made up of six members, including three members from NAE (two from academia and one from industry), two other members from academia, and one from industry. Three of the panel members were also members of the parent committee. The panel was asked to assess the contributions of academic research to integrated logistics services and associated activities, technologies, and methodologies that cut across the many components of the transportation, distribution, and logistics (TDL) services industries. The panel reviewed the literature, developed several case studies, and sent a questionnaire to selected individuals, primarily university-based researchers, with special knowledge of the TDL industry (Addendum). The questionnaire was followed by two roundtables attended by panel members and 12 senior individuals in the TDL services industries (see Addendum).

During the past two decades, deregulation of transportation and rapid advances in computing and communications technologies have resulted in a surge of innovation in logistics and accelerated the pace of change in the broader TDL industries. In manufacturing, reducing inventories and work-in-process through just-in-time deliveries, "pull" systems of supply-chain management, and other technologies and management practices depend on integrated-logistics services, which combine materials management and physical distribution. Logistics has emerged as a distinct function in many companies and as a distinct service performed by integrated-logistics service providers. Over time, the users of integrated-logistics services have become more demanding, and, in response, providers have become more sophisticated. Their use of technology and their need for knowledgeable workers have created interest in and opportunities for academic research.

DEFINITION OF THE INDUSTRY

Transportation, distribution, and logistics services affect every facet of economic life in the United States. In 2000, the United States spent about $1.006 trillion to move freight. Trucking accounted for $481 billion, or about 48 percent of the total. Railroads came in second, accounting for $36 billion, and international, inland, and coastal water transportation was next at $26 billion (Delaney and Wilson, 2001).

As a percentage of gross domestic product (GDP), transportation and inventory costs have been declining since 1981, when total logistics costs in the United States, including inventory carrying costs, transportation, and administrative costs, totaled almost 17 percent of GDP. In 2000, the total was 10.1 percent (Delaney and Wilson, 2001). By far the most important reason for the decline has been deregulation in the rail and trucking industries, but other factors such as de-unionization, advances in technology, and improved management practices have also contributed (Belman and Monaco, 2001). Competition in transportation has not only created new incentives for service providers to reduce costs and improve the quality of service, but has also stimulated innovation in the types of services they provide. At the same time, competition in the retail and manufacturing sectors has forced service providers to reduce costs, ensure faster inventory turns, reduce the amount of work-in-process to a minimum, and operate in close coordination with suppliers. All of these measures have been enabled by improvements in integrated logistics. Figure 5-1 shows the changes in inventory, transportation, and administrative costs in the last 20 years.

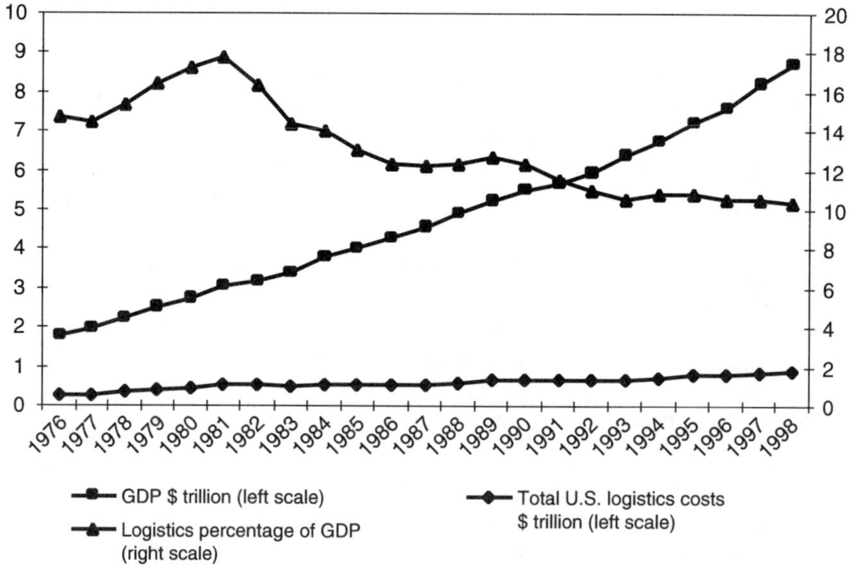

FIGURE 5-1 The cost of logistics in relation to GDP. Source: Cass Information Systems, 1999.

INTEGRATED LOGISTICS: A COMPETITIVE DIFFERENTIATOR

Because of the many changes in technology and management practices in the TDL services industry, especially since deregulation, the panel decided to focus on integrated logistics (also known as supply-chain management) and associated activities, technologies, and methodologies that cut across the TDL services industry. Integrated logistics is becoming an increasingly important source of competition in all parts of the TDL services sector, as well as in other sectors of the economy and the national security establishment.

Two big changes laid the foundation for integrated-logistics services: (1) the automation of transactions enabled by advances in information technology; and (2) the deregulation of transportation. Deregulation has led to dramatically more complex logistics decisions in many industries and a proliferation of transportation service providers, which has encouraged innovation in service delivery and provided businesses with many more choices. Advances in information technology have enabled businesses to accumulate vast amounts of data on every aspect of their supply chains, from production to delivery. Efficient integration of supply, production, and delivery schedules with suppliers and customers requires the effective management of these data.

Supply-chain activities can be categorized into three major areas: (1) the acquisition of materials and supplies; (2) the manufacturing process; and (3) the distribution of products. Supply-chain management is the integration of the flow of materials, documents, information, and finances to optimize individual shipments. Managing the supply chain requires the integration of some parts of the supply chain that were previously regarded as separate.

Integrated logistics includes the planning, implementation, and control of the flow and storage of raw materials, in-process inventory, finished goods, services, related information, and payments among suppliers and consumers from the production of raw materials to the final recycling or disposal of finished goods. The logistics value chain has three major elements:

- the supply chain (the physical components, including manufacturing plants, warehouses, vehicles, and transportation infrastructure)
- logistics business practices (practices and processes associated with the flow of goods, information, and payments through supply chains)
- information and decision technologies (computer-related technologies used to design, plan, and operate supply chains, including the monitoring of the status of materials, parts, and finished products in the supply chain and communications among supply-chain elements)

In a well integrated logistics value chain, all supply-chain elements are optimized with regard to both service and cost. Integrated-logistics technologies can change a “producer-push” system to a “customer-pull” system, in which inventory decisions are based on what customers are actually buying. In manufacturing, this change has been driven by intense competition and the necessity of eliminating waste in the production process in the form of raw materials, work in process, and inventories of finished goods. In retailing, similar pressures to reduce costs, coupled with the growing purchasing power of large retail chains, have led to new ways of doing business. Today, large retail chains choose suppliers based on how well they can match product flow to actual customer demand.

In a survey of major global corporations in 1999, more than 90 percent rated effective supply-chain management as a critical success factor, up from 25 percent in the early 1990s (Deloitte Consulting, 1999a). This change reflects the continuing concentration by businesses on core competencies; increased outsourcing of noncore production, distribution, and other functions; continued emphasis on cost reduction; and product proliferation in the consumer products, food, electronics, and other industries. Integrated-logistics services enable companies to manage the supply chain to meet their cost and flexibility goals.

In the past 15 years, integrated logistics has evolved into a new discipline. Most competitive manufacturing and service companies have installed information systems capable of acquiring large quantities of timely, accurate data regarding major business functions throughout their internal and external supply chains. Advanced planning and optimization (APO) software can respond to the needs of a range of manufacturing systems (Thomas, 1998). Enterprise resource planning (ERP) software, which automates transactions and connects vital business systems (e.g., manufacturing, human resources, financial, and other information/data systems), is also widely used. Industries are attempting to develop optimization and decision-making capabilities that can translate the information generated by these systems into higher productivity and profits. The fruits of academic research have already had a large impact in this area, and they are expected to have a significant impact on the development and application of new decision technologies in support of integrated logistics.

Technology Drivers

Technological, organizational, and contextual changes have significantly influenced performance in integrated-logistics services in the past 10 to 15 years. Significant technological innovations have been focused on applications of information technology:

- hardware that automatically captures data, satellite tracking systems, and navigation systems
- information systems, such as manufacturing resource planning and ERP software, electronic data interchange between firms in the value chain, and database management software
- decision technology

Other innovations include improved transportation, container, and warehouse equipment and improved human-machine interfaces. Contextual changes, which have been both drivers and enablers of change, include: deregulation; the globalization of markets; the emergence of integrated supply chains, just-in-time delivery, and reductions in lot sizes in manufacturing; satellite communication systems; electronic commerce; increased competition; and other general technological advances. The technological foundations of integrated logistics rest primarily on operations research, automated data-capture technologies, and communications/networking technologies. Software providers and third-party logistics providers have been the main drivers of advanced technology development, diffusion, and use in integrated logistics.

Software Providers

Software companies have performed most logistics-related research and development (R&D) (mostly development) in the past 10 years. In manufacturing and logistics, companies such as SAP, PeopleSoft, and Baan have developed a suite of ERP modules for planning production, taking orders, and delivering products. The modules address the following functions:

- production planning (performs capacity planning and creates a daily production schedule for a company's manufacturing plants)
- materials management (controls purchasing of the raw materials needed to build products and manages inventory stocks)
- order entry and processing (automates the data entry process of customer orders and keeps track of the status of orders)
- warehouse management (maintains records of warehoused goods and processes the movement of products through warehouses)
- transportation management (arranges, schedules, and monitors the delivery of products to customers via trucks, trains, and other modes of transportation)
- project management (monitors costs and work schedules on a project-by-project basis)

After nearly a decade of rapid growth and continuous expansion and innovation in product offerings, the ERP market has begun to weaken. The cost and complexity of ERP have created opportunities for other vendors to emerge. By focusing on narrower functionality than ERP and solutions customized by industry, firms such as i2 Technologies and Siebel Systems have rapidly increased their presence in supply-chain and customer management systems (also known as APO solutions systems). For instance, total revenue for Siebel Systems grew from $391.5 million in 1998 to more than $2 billion in 2001(Siebel Systems, 2002).

The emergence of the Internet as a business communication tool has attracted the interest of leading manufacturing, retail, consumer products, and other firms. Despite the benefits of ERP in generating operational improvements, companies continue to feel that their procurement systems face serious challenges and are looking for improvements through business-to-business electronic commerce. In a 1998 survey by Deloitte Consulting, companies reported that 80 percent of their strategic objectives for electronic commerce centered on supply chain and procurement processes. Early on, Ariba and CommerceOne were leaders in electronic procurement, but the large ERP vendors have invested heavily in adding electronic commerce functionality that can be integrated with ERP systems. Although it is still too early to determine which approach or which vendors will succeed, it is clear that large, global firms will invest heavily in online procurement systems (Deloitte Consulting, 1999b).

Third-Party Logistics Providers

The number of third-party logistics providers has increased significantly in the past decade. Third-party providers are companies hired to perform logistics tasks that were previously performed in house. There are several reasons companies decide to outsource planning-intensive functions:

- the explosion of new services in the deregulated transportation provider network
- opportunities provided by new information systems with increasingly sophisticated procedures and automation systems
- the availability of more alternatives
- the volatility of demand
- the rationalization of assets to minimize required investment and maximize return
- the focus on core competencies

Most frequently cited benefits of outsourcing logistics services include (Logistics Best Practices Group, 1997):

- lower costs
- the ability to focus on core businesses
- greater flexibility
- improved expertise/marketing knowledge
- improved customer service

The third-party provider industry has grown significantly in size and scope in the past decade. Revenues grew from $10 billion in 1992 to $56.4 billion in 2000 and were projected to grow 15 percent annually through 2003 (Delaney and

Wilson, 2001). A survey in 1991 of the use of third-party provider services by manufacturers showed that 31 percent of respondents used them; by 1996, 58 percent used them, including about half of the Fortune 500 companies (Logistics Best Practices Group, 1997). Another survey of corporate logistics managers revealed that the outsourcing of the total supply chain to a third-party provider cut logistics costs by more than 20 percent in the first year (Masters and La Londe, 1998). The most frequently used services are warehouse management, logistics information services, shipment consolidation, rate negotiation, fleet management/operations, carrier selection, and product returns (Lieb and Randall, 1996). Information technology, sensor technologies, and communication technologies are essential to all of them.

The hardware and communications technologies that capture the data needed to implement sophisticated ERP and APO software are still relatively new resources among integrated-logistics providers. Despite rapid growth in the industry, a significant learning process is under way as industry leaders learn how to use these tools for competitive advantage, and laggards recognize the inevitability of more technology-intensive business. Among industry leaders, the focus is on the acquisition of information and the effective use of data. Bar coding provides specific identifiers for items being shipped; wireless communications and, increasingly, global positioning systems provide real-time data on vehicle locations. Effective management of these data provides timely information on the location and progress of shipments, which has greatly improved customer service. Perhaps more important, the data can greatly improve efficiency by enabling shippers to match vehicles with excess capacity to nearby shipments headed in the same direction. In the trucking industry, advances in route-planning software have enabled companies to maximize load levels and meet customer expectations (Nagarajan et al., 1999).

The Internet is providing a medium for accelerating and increasing the extent of these changes. In business-to-business electronic commerce, the Internet could lead to even tighter integration of supply chains, and all of the logistics software companies are making their products Internet enabled. As the volume of electronic commerce increases, there is general agreement in the industry that logistics could mean the difference between success and failure. Some of the leading users of integrated logistics (e.g., Cisco Systems and Dell Computer) use the Internet to share information on production schedules, sales, customer orders, stocks, and other critical production data with their suppliers. For instance, as soon as a customer places an order on Dell's web site identifying the specific features being purchased, the order is placed on Dell's production schedule where suppliers can see it and produce the necessary parts. Integrated logistics ensure that the parts arrive at the production line and that inventory for both Dell and its suppliers is at a minimum. The savings have been so large that manufacturers in other industries, from consumer products to automobiles, are exploring ways to apply the Dell model to their supply-chain management.

As Dell has shown, logistics has become a key determinant of success in business-to-consumer electronic commerce. Even if Internet-based retail has little effect on the volume of goods shipped, it will vastly multiply destinations (e.g., individual residences) and raise customer expectations. The result is much more complex logistical systems and an urgency to use technology to manage this complexity profitably. The full implications of the Internet for supply-chain management and the business-to-customer interface are just emerging. Information shared among suppliers is replacing physical inventories as the need for production buffers diminishes. However, different industries will embrace this change at different rates based on a number of factors, such as the number of suppliers and historic relationships in the production chain (Cairncross, 2000).

INNOVATION SYSTEM

Except for software companies and some airlines, very few logistics companies conduct R&D. A few of the leading integrated-logistics service providers conduct a limited amount of internal research and sponsor research at universities. These firms include Schneider National (trucking), United Parcel Service and Federal Express (delivery services), CSX and Union Pacific (railroads), and Sabre/AMR (until 2000) and several major airlines, such as United and USAir (air transport). A few major industrial firms (e.g., Ford, Raytheon, Lucent Technologies, and Procter & Gamble) sponsor research at universities. General Motors, through its Enterprise Laboratory, also supports logistics research. As leaders in their fields, these and other firms play a critical role in the diffusion of advanced logistics technologies.

Most innovations in integrated logistics have come from academic research in transportation/logistics research centers affiliated with university engineering and business schools and from applied research and product development by software companies. Relevant research has also been conducted at national laboratories and transportation centers associated with state departments of transportation.

The academic disciplines involved in research on integrated logistics include applied mathematics, computer science and engineering, industrial engineering, operations research, software engineering, materials science, social and behavioral sciences (human factors), and business and management sciences. Academic research on large-scale optimization models, decomposition methods, integer programming, and network optimization has been extremely valuable to the integrated-logistics industry. Academic research on the phenomenon of electronic commerce, in terms of business models and pricing models, has been crucial to the growth and success of electronic commerce. Economics research on the structure of the industry and its economies of scale were crucial to the debates about deregulation. Business schools involved in logistics research and training tend to focus on the "softer" side of logistics (e.g., management and organization) in contrast to research on software, which is associated with engineering schools.

Transportation research institutes with federal, state, and industry support have been established at some universities to serve as intermediaries between academia and industry and to translate research results to industry. These institutes function primarily as conduits between the academic community and transportation practitioners, adapting technology and research results to meet practitioners' needs and giving them a voice in setting research agendas. Most transportation research centers focus almost exclusively on the movement of people; only a few (e.g., Logistics Institute at the Georgia Institute of Technology, the Center for Transportation Studies at the Massachusetts Institute of Technology [MIT], Stanford University's Global Supply Chain Management Forum, and Princeton University's Computational and Stochastic Transportation Logistics Engineering [CASTLE] Laboratory) are doing a significant volume of work related to moving freight (see Box 5-1).

BOX 5-1
CASTLE Laboratory at Princeton University

CASTLE Laboratory develops and directly implements tactical and real-time optimization models for freight transportation and logistics. Research focuses on the development of general tools for dynamic resource transformation and their adaptation to problems that arise in trucking (truckload [TL] and less-than-truckload [LTL]), rail, and chemical distribution. Tools range from routing and scheduling to fleet management and the forecasting of demand.

CASTLE Laboratory was established formally in 1992 to handle a rapidly growing research program with industry. Annual funding of approximately $750,000 is provided by industry (80 percent) and government (20 percent). Most of the government funds are currently provided by the Air Force Office of Scientific Research. Five corporate partners fund research on optimization models and algorithms for a class of problems called dynamic resource-transformation problems. The techniques are then applied by the sponsoring companies.

Examples of successful techniques developed by CASTLE Laboratory include:

- A network optimization model developed for Yellow Freight and now used by most of the LTL industry has been documented to save $17 million annually (widely believed to be a gross underestimate). This model was instrumental in reducing the number of end of lines in Yellow Freight's network from more than 650 to fewer than 400 with the same geographic coverage.
- A load-matching system developed for TL motor carriers produced a 10 percent reduction in empties when it was first installed.
- A network model for load profitability reduced the operating ratio at Burlington Motor Carriers by 10 percent, restoring the company to profitability.
- A major drayage company with more than 400 trucks uses a real-time scheduling system for routing drivers. Savings are estimated to be more than $10 million annually.

Source: Warren B. Powell, director, CASTLE Laboratory.

A unique academic transportation research center is the Trucking Industry Program (TIP), first established in 1995 at the University of Michigan by a major grant from the Alfred P. Sloan Foundation; the center was relocated to the Georgia Institute of Technology in 2002.[1] TIP contributes to the understanding of the trucking industry through a multidisciplinary approach involving faculty and students from several U.S. universities, including Georgia Tech, the University of Michigan, Michigan State University, Wayne State University, and Duke University. TIP is the only academic program in the United States engaged in comprehensive research on issues associated with labor, the firm, and operations and technology in the trucking industry. TIP is widely known for conducting the most comprehensive survey ever undertaken of truckers at truck stops across the United States. The driver survey, which was conducted between August 1997 and January 1999, has substantially improved our understanding of drivers' work hours and is often cited for presenting the first accurate portrait of truck drivers, their quality of life, and their views of the industry.

Many academic centers require that member companies fund relevant research, provide access to real-world data, and provide sites for implementation. In some cases, access to research results is restricted to member companies, which tend to be the leaders among integrated-logistics service providers and users. At Georgia Tech, member firms pay $50,000 annually to participate, and individual companies are actively involved in research with academic faculty and students. Student/faculty teams work on problems defined by one or more companies, and research results are disseminated actively among members. At Stanford, member firms pay $25,000 to support academic research as part of an industrial consortium. Members benefit from networking and from student/faculty teams working on problems of interest to all participants. At MIT, firms pay $20,000 to share research findings and network but are not directly involved in research.

In addition to research, university transportation research centers also provide executive courses, seminars, and symposia to inform industry of state-of-the-art academic research and new logistics technologies and to inform faculty members of the real problems in industries. For example, affiliate companies of the MIT Center for Transportation Studies come to MIT seven or eight times a year to review the status of academic research.

CONTRIBUTIONS AND IMPACT OF ACADEMIC RESEARCH

Basic research, some of it done in the 1950s with no logistics applications in mind, has had the greatest impact on integrated logistics. Linear programming and integer programming have both made major contributions to methodology (see Box 5-2). Major technological contributions emerged from research on computer science/artificial intelligence, specifically constraint-directed search and its relatives.[2]

BOX 5-2
CPLEX Optimization, Inc.

CPLEX Optimization, Inc., founded in 1987, released its first product, CPLEX 1.0, in 1988. (The company was sold to ILOG in 1997.) CPLEX products are base systems of "simplex" optimizers for linear programming, including a special-purpose network-simplex optimizer. The following add-ons are available:

- mixed integer programming (MIP)
- barrier solver for linear programming
- quadratic solver for convex, quadratic programs
- parallel MIP solver
- parallel barrier solver

All of these optimizers can be considered computational engines that are typically built into other vertical applications.

The company emerged from research on integer programming and combinatorial optimization by Robert Bixby at Rice University. Dr. Bixby wrote the initial versions of what became CPLEX code as a classroom tool in the mid-1980s. The initial impetus for commercialization came shortly thereafter from Tom Baker of Chesapeake Decision Sciences, who began including Bixby's code in his MIMI/LP product as early as 1985.[1] By the time CPLEX Optimization was founded, the code had undergone steady improvements, mainly in response to increasingly difficult, real-world problems.

Simultaneously but separately, fundamental developments were made in the theory of integer programming and combinatorial optimization, motivated by the work of Manfred Padberg (New York University), Ellis Johnson (Georgia Tech), Karla Hoffman (George Mason University), Martin Groetschel (Konrad Zuse Zentrum, Berlin, Germany), and others. As a result, Dr. Bixby recognized a need for a fundamental computational tool that could facilitate the current research in integer programming and its application. The new tool would be a numerically stable, fast, callable, linear programming solver. Subsequent research led to the development of the CPLEX Callable Library, which addressed the so-called traveling salesman problem. Even though the problem itself was the subject of purely academic research with little direct practical applicability, it turned out to be an excellent model for the broad range of functionalities that made CPLEX a successful tool for business applications.

[1]MIMI is a sophisticated commercial system for expressing mathematical models in operations research.

Source: Robert Bixby, professor of computational and applied mathematics, Rice University.

Applied research has also been important to integrated logistics, especially in the areas of large-scale optimization modeling, decomposition methods, network optimization, and other areas of operations research. For example, research at MIT on models for shippers in the logistics industry includes transportation/inventory trade-offs and motor-carrier bidding optimization. Software for routing, production scheduling, and distribution management are examples of high-impact technologies adopted by industry that were developed by university research teams.

Academic research on the development of decision support tools, which require a thorough knowledge of available tools and a thorough knowledge of the industry, has been less transferable. On the one hand, most practitioners cannot explain problems or devise innovative solutions because they are not aware of the available technologies. On the other hand, most researchers are not sufficiently aware of the subtleties of real-world industry problems.

University-Industry Interaction

Modes of interaction between universities and private firms are not industry-specific. Academic research in the TDL industry is disseminated to industry primarily through graduating students entering the workforce who apply what they have learned at the university. Most successful employers recognize this vector of knowledge transfer and try to maximize the expertise of new hires by providing opportunities for them to contribute to changes in company practices.

Other modes of interaction are also important. Member firms in university research centers have a financial stake in the research and an effective interface with researchers. Industry-sponsored research, the commitment by one or a small number of companies to support a specific project, is also an effective method of generating research that benefits industry. Companies are likely to adopt the research results of projects they have helped develop and funded (or partly funded).

Consulting arrangements for faculty are another method of moving research results into the field. Although the relationship between consulting and technology transfer is not well documented, faculty consulting provides an obvious mechanism for generating new practices in industry. It also provides faculty with much needed exposure to industry problems, which has enormous benefits in shifting research from interesting but theoretical subjects to useful and applicable subjects. In logistics, academic consulting has often been a precursor, as well as a complement, to academically originated software start-up companies. For instance, start-up companies (predominantly software companies) in decision technology, founded by professors and based on their academic research and consulting, have made significant contributions to innovation in integrated logistics. The special role of spin-off companies is described below, and an example is given in Box 5-3.

Presentations at conferences with extensive industry participation, such as the conferences sponsored by the major operations-research/logistics-management professional societies (e.g., Institute for Operations Research and Management Science, the Council on Logistics Management, the International Society of Logistics), and other organizations are also an important vector for communicating academic research results to industry.

BOX 5-3
Princeton Transportation Consulting Group

Princeton Transportation Consulting Group (PTCG) was founded in 1987 by Drs. Cape, Powell, and Sheffi, faculty members at Princeton University and the Massachusetts Institue of Technology.[1] PTCG was started by implementing Drs. Powell and Sheffi's consulting practice, which was based on their university research. The company, which provides decision support systems for truckload (TL) and less-than-truckload (LTL) carriers, is the leading provider of real-time TL dispatching software and LTL load-planning software to the motor carrier industry. It also provides systems to the logistics industry for inbound transportation with multiple plants and distribution centers and optimization-based bidding software. This software was developed from Dr. Sheffi's research at MIT and was inspired by his experience with bidding transportation services through LogiCorp. Based on experience, he was able to identify a new category of optimization software and develop a new approach. The solution was later developed into an application by PTCG and is now used by dozens of large manufacturers and retailers throughout the United States, Europe, and the Far East.

[1]In 1996, the company was acquired by Sabre (at the time a subsidiary of AMR, owner of American Airlines) and became its Boston office. In 2000, Dr. Sheffi bought Sabre's logistics assets to form Logistics.com.

Source: Yossi Sheffi, codirector, Center for Transportation and Logistics, Massachusetts Institute of Technology

As professional education in logistics management, in the form of short courses, increases, it could encourage the use of, and demand for, new logistics tools. The technology transfer via these courses is generally minimal. They expose practitioners to new technologies and ideas but generally provide little information on how to use them effectively.

All of these mechanisms of knowledge transfer show that closer ties between companies and universities are necessary for industry to reap the maximum benefit from academic research. Programs like MIT's Corporate Affiliates Program and Georgia Tech's Leaders in Logistics are reasonably good avenues of knowledge transfer because they provide for continuity of participation and a long-term learning process for faculty as well as industry. The panel believes that many faculty members focus on solving nonexistent problems, not because they are poor researchers, but because they do not understand the real-world problems faced by industry. This problem can usually be overcome if faculty members are given sufficient opportunities to interact with industry.

Spin-off Companies

The creation of new companies based on research results is an important mechanism for commercializing academic research results and increasing their impact on industry. The case studies prepared for this report describe companies started by university professors; these are primarily software companies for implementing products or services developed through research and consulting.[3] One of the case histories is summarized in Box 5-4.

In the field of integrated logistics, most start-up companies involve decision support systems (routing and scheduling), optimization software, traffic network analysis, third-party logistics companies, and consulting services. Georgia Tech's Logistics Institute, MIT's Center for Transportation Studies,

BOX 5-4
AD OPT Technologies, Inc.

Headquartered in Montréal, Canada, AD OPT Technologies was founded in 1987 by professor François Soumis (École Polytechnique de Montréal) and four programmers from two university research centers in Montréal: Groupe d'études et de recherche en analyse des decisions (GERAD) and Centre de recherche sur le transport (CRT). AD OPT currently employs 100 people and had revenues of $15 million in 2001.

As a spin-off of GERAD and CRT, AD OPT is an example of the commercialization of university research results. The company continues to maintain close ties with universities in Montréal through personnel, sponsored research, and graduate hirings. François Soumis of École Polytechnique de Montréal, a cofounder of AD OPT, and Jacques Desrosiers of École des Hautes Études Commerciales, who joined AD OPT in 1992, supervise the GENCOL team at GERAD. GENCOL, an acronym for *GÉNération de COLonnes*, embodies the principles of the column-generation solution method that was developed to solve large-scale complex problems. The system can be modeled so that solutions correspond to the set of paths found in vehicle routing and crew scheduling. Since GENCOL's conception in 1981, it has evolved continually through the research of many graduate students and computer analysts. This program is the result of an enormous amount of work funded largely by grants and industry contracts. The sale of products based on the GENCOL optimizer represent more than 60 percent of AD OPT's turnover. The present and future growth of the company are directly related to research at GERAD.

Initially AD OPT Technologies provided consulting services to optimize operations of open-pit mines and the routing of propane delivery trucks. In the early 1990s, the company began to focus on product development and scheduling of airline crews. The *ALTITUDE* Crew Pairing system was the first product developed and marketed by AD OPT. Crew rostering (Preferential Bidding and Bidline) systems followed. *Shift*Logic, a shift scheduling and management tool, resulted from work done for NavCanada on a tool for scheduling air traffic controllers throughout Canada. Today, AD OPT continues to invest heavily in research and development to maintain its leadership and technological advantage in airline and personnel scheduling systems for air traffic control and other vertical markets, such as ambulance, police, firemen, 911, nurses, and doctors.

Source: Jacques Desrosiers, professor, Department of Management Sciences, and member of the Group for Research on Decision Analysis, Ecole des HEC Montréal.

and Princeton's CASTLE Laboratory have served as incubators/support networks for start-up founders, who received feedback from other university researchers and/or companies working on real-world logistics issues and problems (see Box 5-5). Once a new company is created, graduate students are usually hired to pursue further development of software products. Based on the accumulated knowledge, problem-solving skills, and expertise of professors and students, the software is tailored to solve a customer's problems. Considering the success of many of these start-up companies, the impact on industry of academic research commercialized in this manner has been substantial. In many cases, larger companies eventually acquire the start-up companies, thus providing more resources for continued product development and more extensive marketing.

BOX 5-5
CAPS Logistics

CAPS Logistics was founded in 1979 by Don Ratliff and John Jarvis, professors of industrial and systems engineering at Georgia Institute of Technology. The company provides three families of decision-support software for supply-chain logistics: supply-chain design and coordination, shipment planning, and dedicated fleet management. CAPS has 120 full-time employees; revenue totaled more than $15 million in 1998, when the company was sold to Dutch-based enterprise software giant Baan.

Much of the methodology implemented in CAPS software was developed through academic research. Many of the concepts for integrating interactive map-based graphics with network optimization emerged from research at Georgia Tech. The network optimization concepts were developed at a variety of universities, including Georgia Tech and MIT. The research at Georgia Tech was funded by a combination of military funding agencies, NSF, and private companies participating in the Leaders in Logistics Program. CAPS software would not have been possible without this research.

CAPS also has benefited directly from federal support for logistics research. In 1985, CAPS received a Phase 1, and later a Phase 2, Small Business Innovation Research (SBIR) grant from the Office of Naval Research to develop concepts for a "logistics tool kit" that could be used to develop flexible logistics software. This became the CAPS Logistics Toolkit, which evolved into the foundation for CAPS software. CAPS received six other SBIR Phase 1 grants but, because of matching requirements, particularly from NSF, was not able to obtain Phase 2 funding for any of them. Therefore, additional development and transfer of the technology had to be funded internally. This research was focused primarily on making the technology easy to use by nontechnical people. CAPS currently has more than 1,000 software installations. The users are primarily Fortune 200 companies.

Source: Donald Ratliff, Regents' Professor and UPS Professor of Logistics, and director, Logistics Institute, Georgia Institute of Technology.

Sources of Funding

Both industry and government provide funding for academic research in logistics, although federal funding for basic research has decreased recently. The U.S. Department of Defense has historically supported most of the basic research relevant to integrated logistics; as defense budgets have decreased, these funds have become increasingly difficult to obtain. The Defense Advanced Research Projects Agency (DARPA), which funds a considerable amount of applied research in logistics, directs most of its funding to consulting firms and very little to academia. The Defense Logistics Agency (DLA), which is a member organization of university logistics centers, such as the Logistics Institute at the University of Arkansas, provides limited funding for academic research, as does the Federal Aviation Administration (FAA). For example, the FAA funds the National Center of Excellence for Aviation Operations Research at the University of California at Berkeley and MIT.

Historically, the National Science Foundation (NSF) has funded research relevant to logistics in mathematics and industrial engineering, but logistics-related research has not been a priority. In 2001, however, NSF created the Center for Engineering Logistics and Distribution (CELDi), a multiuniversity, multidisciplinary industry/university cooperative research center based at the University of Arkansas, University of Oklahoma, University of Louisville, and Oklahoma State University. Research is driven by, and sponsored by, member organizations, which include manufacturing, maintenance, distribution, transportation, information technology, and consulting companies. CELDi emerged in 2001 from the Material Handling Research Center (founded in 1982) and the Logistics Institute at the University of Arkansas (established in 1994) to provide integrated solutions to logistics problems through research using modeling, analysis, and intelligent systems technologies.

Perhaps the most significant source of federal funding in TDL research comes from the U.S. Department of Transportation (DOT) through its University Transportation Centers (UTC) Program. The Transportation Equity Act for the 21st Century (P.L. 105-178) of 1998 authorized up to $194.8 million for grants to establish and operate as many as 33 UTCs throughout the United States. In addition to emphasizing the educational role of universities, the program funds basic and applied research to advance the body of knowledge in transportation. All UTCs are required to match federal funds dollar for dollar; state departments of transportation typically provide the match. Some UTC focus areas are listed below:

- intelligent transportation systems
- advanced technologies in transportation operations and management
- advanced infrastructure and transportation
- advanced transportation simulation
- advanced transportation technology

Industry funding of TDL research is limited and concentrated in a handful of the biggest logistics research centers. Projects funded by industry, which can be either proprietary or generic (i.e., results can be published), cover a wide range of topics, from the movement and tracking of material in a factory to the distribution of finished goods to global markets to the scheduling of aircraft and crews for airlines. (The case studies provide some insight into the commercial applications of logistics technologies and research conducted to refine solutions to commercial problems.)

Impediments

Although the contribution of academic research to the emergence of sophisticated integrated logistics management and optimization tools has been significant, and the resulting decreases in operating costs and improvements in efficiency have been high, very little public or private funding is available for fundamental research in logistics. A critical mass of funding tends to be concentrated in academic logistics institutes where member companies help to define and then participate in research projects. However, industry participation sometimes creates tensions between academic researchers and industry managers who may have conflicting objectives for collaborative projects. Industry wants timely solutions to practical problems, while academics want to conduct valid, preferably "novel," research with publishable results. For instance, stable, easy-to-use logistics software tools may be a high priority for industry sponsors but may be less appealing to researchers than refining and perfecting existing tools or developing new tools.

An infrastructure for logistics research has not been fully developed. Diagnostic tools for the complex software used in logistics applications have only recently emerged. Because there are no common data sets, data must be gathered from industry for each project. If the data are considered proprietary, a company may refuse to provide them. Libraries of test data sets on small mathematical problems have been developed, but not on large problems of logistics. As a result, researchers often do not have access to data, and certainly not consistent data, on which to base experiments.

Areas for Future Research

In general, the TDL industry is becoming more sophisticated. Continued progress in the implementation of ERP and related software tools and the development of a stronger information technology infrastructure will result in the accumulation of better and more extensive data, which, in turn, will enable the development and implementation of more sophisticated decision-support tools, which are likely to be developed through academic research. Universities might also assume a larger role in the training/retraining of technically sophisticated logistics managers in the use of new tools.

Academic research might have a larger impact in the area of linkages between manufacturing and transportation, especially as companies continue to try to reduce inventories, transport costs, and time through integrated supply-chain management. Another fruitful area for research would be the development and use of optimization procedures for the macrolevel coordination of the supply chain, such as the allocation and scheduling of activities among plants, warehouses, and transportation channels. For example, a production schedule that would mean low manufacturing costs might increase transportation costs. Because these costs are incurred (and measured) by different departments, manufacturers may not take transport costs into consideration in their planning. As long as companies had buffers (time and/or inventory) to decouple these processes, this was not much of a problem, but it will certainly become a major problem soon. Forecasting transportation demand is becoming increasingly important, and capacity planning by carriers will have to be even more rigorous to account for uncertainties/variabilities in demand.

University research could play a larger role in other areas, such as the integration of planning and operations, the development of information technology to support order placement, the development of information technology to support the coordination and scheduling of the movement of goods, and consumer research. For instance, with the availability of real-time information (e.g., current inventory levels at retail stores; current traffic information), models of sequential decision making under uncertainty may become a basis for logistics tools in the future. Research in these areas would be "engineering research" (i.e., research focused on bringing technologies to bear on industry problems for which few principles or laws exist). Transportation and logistics security has become a critical issue that will also pose important research questions in the future: for example, how to improve the security of systems while continuing to improve, or at least maintain, their efficiency.

FINDINGS

Finding 5-1. The contributions of academic research to integrated logistics have been significant in areas of basic research as well as in the development and application of specific software technologies. However, academic research on logistics and technologies has had a moderate impact on the transportation, distribution, and logistics industry overall.

Except for the involvement of member companies in academic logistics-research institutes, no institutionalized methods of technology transfer from universities to the TDL industry have been established. Most high-impact academic research in operations and decision-support sciences can be attributed to faculty members who either translated their research findings into software commercialized through a start-up company or who were closely involved with a specific company in the industry. For the latter route to be effective, the host company must be sophisticated enough to be receptive to innovative research, and the researcher must be familiar with the specific needs of the company. Industries that have been most influenced by academic research are airlines, manufacturers that operate private truck fleets, less-than-truckload and truckload motor carriers, and software companies.

Without a strong tie to industry, academic research tends to be disconnected to the needs of industry. The academic imperative that researchers publish their work often creates barriers to their addressing real industrial needs using real industrial data. Research problems that could lead to publishable results may be far too complex for companies to benefit from the results. For these reasons, the commercialization of results through start-up companies may continue to be the most effective pathway for academic research to affect industry performance.

Finding 5-2. Federal funding agencies have not recognized logistics as a separate intellectual discipline and thus have not funded long-range, potentially high-impact academic research in the field.

The DOT, primarily through the University Transportation Centers Program, funds some research relevant to logistics. NSF provides some funds for programs that bring industry and academia together in the logistics area (e.g., CELDi, the multiuniversity industry-university cooperative research center formed in 2001). In addition, NSF has launched a modest initiative, Exploratory Research on Engineering in the Transport Industries, that specifically addresses logistics and supply-chain management. Although some NSF-funded research has benefited airlines that contributed matching funds, NSF's funding for research in logistics has been minimal. In the past DOD (particularly the Office of Naval Research and the Air Force Office of Scientific Research) funded logistics as a discipline, but the funds are now focused more on mathematics research. DARPA has an extensive applied research program in logistics, but funding has been directed mostly toward consulting companies rather than universities.

Finding 5-3. The two most influential factors on the industry are (1) the explosive growth of information technology and complementary disciplines and (2) deregulation of the freight industries.

Finding 5-4. Most companies are still in the process of adopting and implementing information systems and acquiring data from the new systems. The next wave of innovation will be to apply the data to optimization models to support decision-making capabilities.

Decision-support software can be divided into two categories: (1) the automation of transactions to increase productivity and improve quality, and, in the process, create and collect data; and (2) the optimization of management decisions based on these data. Most companies have been in the information-acquisition stage for some time and have accumulated large quantities of accurate, real-time data. Optimization and decision-making support tools could be used to translate these data into information that could lead to higher profits and competitive advantage.

Finding 5-5. Except for airlines, few companies in the TDL industry conduct their own research, and those that do are primarily interested in development rather than research.

Airlines have not only invested in their own research, but have also demonstrated their support for academic research by funding research projects and participating in logistics-research institutes. Most other TDL firms have not yet realized the value of R&D for increasing efficiency and establishing a competitive advantage. These companies have a long history of adopting low-technology, manual-intensive business models. Even the competitive demands of the industry have not yet created a significant demand for research. Most research, therefore, has influenced them through software developments, either by existing software firms or new start-up companies.

Finding 5-6. Providers of logistics software are undergoing rapid consolidation.

Consolidation in the logistics software industry has narrowed the range of industry-research interfaces resulting in fewer spin-off companies. Consolidations may eventually create economic units large enough to support more basic research.

Finding 5-7. The lack of industry demand for research and research-based innovations has made it difficult for academic researchers to identify useful, high-priority problems.

The absence of organized interactions between academia and industry reflects the general lack of an organized innovation system in the logistics industry. Most academic researchers have little incentive to work with industry or to understand industry issues. Because NSF and other funding agencies do not recognize logistics as a discipline, academics are reluctant to go into the field. The lack of interest on the part of industry, particularly the lack of corporate R&D departments to provide an interface with researchers, reinforces this reluctance.

Finding 5-8. Research important to industry requires researchers who understand the industry and the subtleties of the problems industry faces.

The experience of company-supported logistics institutes has demonstrated the importance of having a critical mass of academic researchers who thoroughly understand both the problems industry faces and the academic methodologies needed to solve them. Although these research subjects meet rigorous academic criteria and address the real needs of industry, much more work will have to be done to encourage industry interest in academic research.

Finding 5-9. Transportation-research institutes have the potential to improve industry-university interaction significantly. MIT's Center for Transportation Research, Georgia Tech's Logistics Institute, Stanford's Global Supply Chain Management Forum, and Princeton's CASTLE Laboratory have shown that they can improve the transfer of information between academia and industry.

Historically, these four centers have accounted for most industry-related logistics research in academia; all four rely on industry participation and funding. With NSF support, CELDi continues this model, requiring industry participation but in a multiuniversity context. The effectiveness of these institutes is a function of their critical mass of academic researchers who can interact effectively with industry. The continuity of industry participation in research, outreach, networking, and continuing education has initiated a two-way learning process.

RECOMMENDATIONS

The flow of people between academic research and industry must be greatly increased. The panel recommends that the following steps be taken to further this end.

Recommendation 5-1. Academia should develop better curricula and programs in logistics to attract student, faculty, and industry interest. This will require changes in the incentive and reward systems to encourage qualified researchers to work in the field.

Recommendation 5-2. Industry should establish sabbaticals in industry, full-time or part-time teaching by industry practitioners in universities, and other programs that would promote university-industry interactions.

NOTES

[1]TIP moved in 2002 when the director, Dr. Chelsea C. White, joined the faculty of the School of Industrial and Systems Engineering (ISyE) at the Georgia Institute of Technology.

[2]Constraint-directed search is one of a class of constraint-satisfaction problems in artificial intelligence in which knowledge is expressed declaratively as a collection of explicit, categorical constraints over a set of possibilities.

[3]Companies profiled include LogiCorp, Inc., PTCG, Inc., CAPS Logistics, CPLEX Optimization, Inc., AD OPT Technologies, Inc., Transport Dynamics, Inc., STS, and Cambridge Systematics, Inc.

REFERENCES

Belman, D., and K. Monaco. 2001. The effects of deregulation, de-unionization, technology and human capital on the work and work lives of truck drivers. Industrial and Labor Relations Review 54(2A): 502–524.

Cairncross, F. 2000. Inside the machine: a survey of e-management. The Economist, Supplement 357(8196): 5–40.

Cass Information Systems. 1999. Wall Street's View of Logistics. Annual Session of the Council of Logistics Management, Toronto, Ontario, October 18, 1999.

Delaney, R., and R. Wilson. 2001. Managing Logistics in a Perfect Storm. 12th Annual State of Logistics Report. Washington, D.C.: Cass Information Systems.

Deloitte Consulting. 1999a. Energizing the Supply Chain: Trends and Issues in Supply Chain Management. New York: Deloitte Consulting.

Deloitte Consulting. 1999b. Leveraging the e-Business Marketplace. New York: Deloitte Consulting.

Lieb, R.C., and H. Randall. 1996. A Comparison of the Use of Third-Party Logistics Services by Large American Manufacturers, 1991, 1994, 1995, and 1996. Lexington, Mass.: Mercer Management Consulting.

Logistics Best Practices Group. 1997. Survey on Logistics Outsourcing Practices. College Park, Md.: University of Maryland Press.

Masters, J., and B. La Londe. 1998. The 1997 Ohio State University Survey of Career Patterns in Logistics. Columbus, Ohio: Ohio State University Press.

Nagarajan, A., J.L. Bander, and C.C. White III. 1999. Trucking. Pp. 123–153 in U.S. Industry in 2000: Studies in Competitive Performance. Washington, D.C.: National Academy Press.

Siebel Systems, Inc. 2002. Annual Report 2002. Available online at: *http://www.siebel.com/common/includes/pdf_frame.shtm?pdfurl=/downloads/about/pdf/siebel_2002_annualrpt.pdf.*

Thomas, J. 1998. The Quest Continues. Logistics Management and Distribution Report 37(8): 39–41.

ADDENDUM

Questionnaire

The following questionnaire was sent to selected individuals, primarily university-based researchers, with special knowledge of the transportation, distribution, and logistics industry. Included among the questionnaire respondents were a senior executive at the Sabre Group and professors with expertise in operations research, applied mathematics, industrial engineering, transportation research, and integrated logistics from Georgia Institute of Technology, Massachusetts Institute of Technology, Northwestern University, Princeton University, University of Maryland, and Universite de Montréal.

IMPACT OF ACADEMIC RESEARCH ON INDUSTRIAL PERFORMANCE

1. Briefly describe research carried out in your university that has had an impact on integrated chain logistics within the transportation, distribution, and logistics services (TDL) industry.[1] If possible, please provide information about the nature of specific research contributions; the level of faculty, student, and industrial researcher involvement; the research time frame; sources and level of funding; how research results were transferred (specific steps) to the industry, and whether the research had implications for companies and industries beyond the specific company or industry involved.

2. Please describe briefly any other academic research they you believe has had a notable impact on the TDLS industry. (See suggested issues/items in subtext to question #1.)

3. Overall, would you describe the impact of academic research on industrial performance in the TDLS industry as (circle one):

   1. very large
   2. large
   3. medium
   4. small
   5. very small/non-existent

[1]The NAE panel defines the TDLS industry as encompassing all businesses involved in the transportation and storage of goods and the movement of people. Its constituent parts include: carriers, third party logistics firms, management consulting firms, terminal and distribution firms, warehousing companies, shipping companies, software providers, and major customers/retailers. The focus of the NAE panel study, however, is on the contributions of academic research to integrated chain logistics and associated activities, technologies, and methodologies that cut across the TDLS industry's many constituent parts.

4. Briefly describe significant emerging trends or developments in the TDLS industry in which university research could or should play a larger role.

5. What are the most important actions universities could take to enhance the contributions of academic research to performance in the industry?

6. What are the most important actions companies could take to enhance the contributions of academic research to performance in the industry?

7. What are the most important actions government could take to enhance the contributions of academic research to performance in the industry?

8. What is your estimate of total annual research dollars spent at your institution on academic research related to the TDLS industry? What shares of the total would you estimate are funded by government, industry, or other sources?

9. Other comments?

Industrial Roundtable Agendas and Participant Lists

INFORMATION TECHNOLOGY IN THE LOGISTICS, TRANSPORTATION, AND DISTRIBUTION INDUSTRY A ROUNDTABLE DISCUSSION

October 12, 1998
Radisson Hotel, Maingate Anaheim
Anaheim, California

Roundtable Agenda

6:00 **Welcome**
Chris Lofgren, Schneider National, Inc.

Introduction
Don Ratliff, Georgia Institute of Technology

6:10 **Roundtable Participants**
Self-introduction; major technology needs and trends in the workplace

7:10 **Group Discussion**
Decision Support Systems: current and future trends and needs

8:00 **Session Wrap-Up and Adjourn**
Chris Lofgren

Participants

Industry Experts
Chair: Chris Lofgren, Chief Technical Officer, Schneider National, Inc.
Christopher Caplice, Senior Consultant, The Sabre Group
Virginia Carmon, Senior Manager, KPMG Peat Marwick LLP
Bernard Hale, Senior Vice President Customer Support, DSC Logistics
Thomas Sanderson, President, Sabre Decision Technologies
Jay Mabe, Associate Partner, Andersen Consulting
John Coyle, Kimberly Clark Corporation

NAE Panel Members:
H. Donald Ratliff, Regents Professor and UPS Professor of Logistics and Director, Logistics Institute, Georgia Institute of Technology

Cynthia Barnhart, Associate Professor, Civil and Environmental Engineering Department, Massachusetts Institute of Technology
Robert E. Bixby, Professor, Department of Computational and Applied Mathematics, Rice University

NAE Staff
Diane Alberts, NAE Fellow, Program Office
Proctor Reid, Associate Director, Program Office
Nathan Kahl, Project Assistant, Program Office

CHALLENGES AND TRENDS IN THE LOGISTICS, TRANSPORTATION AND DISTRIBUTION INDUSTRY A ROUNDTABLE DISCUSSION

October 13, 1998
Radisson Hotel, Maingate Anaheim
Anaheim, California

Roundtable Agenda

Welcome and Opening Remarks
Professor H. Don Ratliff (Georgia Institute of Technology) and Professor Cynthia Barnhart (Massachusetts Institute of Technology)

Self-introduction of Roundtable Participants:
Describe your job and three major challenges that you currently face.
Specific questions to focus your thoughts:

1. What are your thoughts on the changing nature of the logistics business?
2. What are the current and future technology needs of your business?
3. Name the most important technological innovation in your business in the recent past?
4. Do you interact with / turn to colleagues at universities to help meet your business challenges? If so, how? If not, why not?
5. What is an important emerging trend in the LTD industry?
6. What are your human capital needs? What are challenges?

List of Participants
Industry Experts:
Don Schneider (chair), President, Schneider National, Inc.
Doug Duszynski, Director of Transportation, The Quaker Oats Company

Wayne Power, Director, Global Logistics, Owens Corning
Vince Chiodo, Director of Supply Chain Management, CHEP USA
Hank Dehne, Manager of Logistics Distribution & Planning, General Mills

NAE Panel Members:
H. Donald Ratliff, Regents Professor and UPS Professor of Logistics and Director, Logistics Institute, Georgia Institute of Technology
Cynthia Barnhart, Associate Professor, Civil and Environmental Engineering Department, Massachusetts Institute of Technology
Robert E. Bixby, Professor, Department of Computational and Applied Mathematics, Rice University

NAE Program Office Staff:
Proctor Reid, Associate Director
Diane Alberts, NAE J. Herbert Holloman Fellow
Nathan Kahl, Project Assistant

6

Report of the Panel on the Financial Services Industry

The Panel on the Financial Services Industry comprised six members, including one NAE member from industry, two other members from industry, and three members from academia. Three of the panel members were also members of the parent committee. The panel reviewed the literature and sent a questionnaire to experts in academia, the financial services industry, and government. The questionnaire was followed by a workshop attended by 21 senior individuals in the financial services sector (see Addendum).

Financial services are the foundation of a modern economy. They provide mechanisms for assigning value, exchanging payment, and determining and distributing risk, and they provide the essential underpinnings of global economic activity. The financial services industry encompasses banking, insurance, equity markets and brokerage houses, leasing, venture capital, and personal and commercial credit. The industry provides the wherewithal for the capital investment that drives innovation and productivity growth throughout the economy. As the economy has become larger, more global, and more complex, the industry has become increasingly innovative in delivering products and services.

The scope of the financial services industry is enormous:

- U.S. financial institutions process more than $9 trillion per day, approximately equivalent to the U.S. gross domestic product (GDP) in 2001.
- Daily average trading volume on the New York Stock Exchange was more than one billion shares in 2000.
- Commercial banks in the United States hold more than $6 trillion in assets (Federal Reserve, 2002).

- The annual global life insurance market is more than $1.2 trillion in written annual premiums; the United States accounts for 24 percent of this total. Other kinds of insurance total more than $900 billion in annual policies; the United States accounts for 40 percent.
- The U.S. savings and loan industry holds more than $1 trillion in assets.
- The industry employed nearly 6 million people in 2000 (U.S. Bureau of the Census, 2002).

Financial services has long been perceived as a staid, conservative industry. However, the growth of information technology and global competition has driven the industry to embrace rapid technological advances and innovation. Information technologies, the Internet, risk modeling, and the creation of new markets in derivatives and other financial instruments have accelerated the pace and volume and increased the efficiency of financial transactions, at the same time driving the industry to make significant investments in new technologies. For instance, financial services firms account for more than 60 percent of investments in information technology in the United States. Hence, highly technically trained experts have become increasingly important to the industry.

These trends have also led to an increase in competition and a blurring of the definition of financial institutions. Financial services firms are involved in leasing (GE Capital Corporation is perhaps the leading example), personal credit (e.g., AT&T's entry into and subsequent exit from the credit card business)[1] and other activities that were once the preserve of commercial banks. With the virtual elimination of interstate banking restrictions and mergers of firms from disparate parts of the industry, consolidations have proliferated in the banking industry. The end of the Glass-Steagall Act of 1933, which prohibited banks from accepting deposits and underwriting securities and established the Federal Deposit Insurance Corporation, and the constant entry of new domestic and global competitors, have created an industry subject to continuous redefinition and restructuring. Technologies and services for institutions and individual investors on a national and global scale are now widely available.

The consequences of rapid innovation, continuous restructuring, and ongoing regulatory changes have been largely beneficial to the U.S. economy and society. However, these trends have also had some negative consequences. Some innovations have been credited with having a destabilizing effect on financial markets leading to "failures" that have been very costly to the industry in both dollars and public trust. For example, the application of derivative pricing formulas (developed by academics) in the large global hedge fund, Long Term Capital Management, was implicated in the fund's near collapse in 1998.

DEFINITION OF THE INDUSTRY

Because new players have entered the field from areas not traditionally associated with the industry, and firms already in the industry are expanding their services to encompass different parts of the business, a narrow definition would not be useful. The adoption of new technologies to provide new services and existing services in innovative, more convenient ways further complicates efforts to define the boundaries of the industry. Defining the industry using specific companies or current participants as examples may also be somewhat misleading.

Therefore, the panel decided to approach a definition from two directions. The first is based on a functional view of financial services, which is perhaps the most useful way to examine how and why the industry is changing. In the functional view, most financial services or processes can be divided into two broad categories: (1) the packaging and description of securities; and (2) the facilitation of trade. Each function poses very different questions for academic researchers.

The second approach is to examine changes in the industry over time since about 1980, when it still resembled the traditional industry, which was dominated by commercial and investment banks, insurance companies, and brokerage houses. Transitions in the industry can then be assessed by comparing the industry in 2002 with the 1980 version. The characteristics of the industry in the near future can be suggested based on current trends. This approach provides a sense of the changes taking place in the industry over time.

Functional View

The first function of the financial services industry is the packaging and description of securities for consumers and businesses. This function spans a wide range of core services, from the mundane, such as providing savings accounts, certificates of deposit, and insurance for consumers, to the esoteric repackaging of risk for large corporations, such as security bundling, (e.g., mutual funds and mortgages). Recently, repackaging services have been marketed like real goods that are manufactured by a factory and brought to the attention of consumers by a marketer. In the complex world of finance, the industry must also play an advisory role. In fact, many firms focus more on advising investors than on selling securities, although the recent weakness in the stock markets and the wave of auditing scandals have made this function more difficult.

The second major function of financial services is facilitating trades, or helping implement transactions. Financial service firms have always played a key role in moving trades involving actual goods and services or financial products through the economy. A simple example of this type of activity is the interest-free checking account: the bank provides standard check-clearing functions to help consumers and businesses make trades but does modify the securities that are traded. Other examples of transactional services are automated teller machines (ATMs), foreign exchange, and debit cards.

Changing Nature of Financial Services

Until 1980, the financial services industry could be divided into four distinct categories: commercial banking, investment banking, insurance, and investment management. Applying the functional definitions to these categories is relatively easy. Commercial banking served both functions, but the other three primarily packaged securities and advised customers. Full-service brokerage houses performed advisory and transactional roles.

The financial services industry in 2002 is in a state of continuous change, and pinning down exactly what is happening is very difficult. Many outside forces are at work, including macroeconomic changes and globalization, technological innovation, competitive pressures, evolving regulatory climates, and specific events in the market. Given that financial services are really almost entirely information, the most powerful force for change is in information technology, especially the emergence of networks, such as proprietary information networks that deliver up-to-the-minute financial information, greatly facilitating financial transactions. The astounding impact of advances in information technology is immediately apparent when the amount of money moved, the number of shares traded, or the speed of transactions are compared with the same figures for the early 1980s.

Technological changes are intertwined with other forces, especially changes in domestic and international regulation, which are transforming many parts of the industry. The number of traditional companies is decreasing because of consolidations, which are also bringing firms together from disparate parts of the industry to form new entities. The increase in global linkages means that few financial service firms are immune to macroeconomic events that spill over national borders. Both of these trends—consolidation and globalization—have raised questions about the stability and safety of traditional financial services firms and the ability of regulators to oversee them.

Because of changes in industry structure, regulatory environments, the scope of services, and technological capabilities, the boundaries of the current financial services industry are very difficult to define, and may be even more difficult to define in the future. New firms that are not in the conventional circle of financial services are providing traditional functions. Examples can be found in e-commerce, where new entrants include firms that heretofore were associated with computer software and hardware and entirely new web-based start-up companies. Both of these transaction and information service providers are outside the traditional realm of commercial banking but provide many of the same services. Note, however, that by regulation, commercial banks are the ultimate providers of money in all forms. A similar situation has arisen with new and potential providers of electronic money.

Electronic commerce is not the only area pushing the boundaries of financial services. The bundling and repackaging of securities is more extensive than ever before in areas that are not obviously related to financial products. For instance, new musical recordings are "securitized" to spread the risk if sales of a recording are poor and the rewards if the recording is a hit. Other kinds of securities also blur the boundaries. An investment bank might issue options on physical events, such as earthquakes, which have been the province of insurance companies. The growing areas of equipment leasing and risk management are also difficult to classify; they include many services provided by actuaries, accountants, and other consultants whose activities are difficult to classify along traditional lines.

The industry today includes many nontraditional functions, such as retailer sales, software development, network management, and key functions in the consumer industry. In general, these innovations have radically changed the key financial players. Many firms whose primary line of business is unrelated to financial services are integrating financial capabilities into their traditional business offerings to provide innovative business approaches. This trend is especially prevalent in the world of e-commerce and Internet business. Also, core parts of the industry, such as consumer banking, are now facing competition from companies that historically had nothing to do with financial services.

As consumers are increasingly able to make investment decisions themselves, the industry is being forced to adapt. The government and others have been encouraging people to take more control of their long-term financial needs, and consumers are now more active in their long-range investment planning. The Internet has enabled firms to deliver detailed financial information and advice directly to consumers' homes. Indeed, unlike the old "transaction-based" relationships, financial advice has become a key factor in the establishment of enduring relationships with clients. This combination has opened new areas of investment advising, and new players have entered the field, ranging from simple web sites that provide stock quotes to complete data and software services provided by firms like Quicken and Microsoft.[2]

The shape of the industry in 5 to 10 years is difficult to predict. Change is likely to continue, and the industry is likely to become increasingly commodity oriented with an emphasis on the development of relationship-based interactions with customers. Future participants are likely to have more direct contact with consumers, who, in turn, will have more choices to make. To meet the needs of consumers, the industry is likely to be under great pressure to use more sophisticated marketing approaches. Financial services are likely to be provided in chunks corresponding to certain functions, and overlaps with other services are likely to decrease. Crucial people in the industry are likely to have technological backgrounds rather than traditional commercial or investment banking backgrounds. Future industry extensions will be difficult to classify along conventional business lines. Some will fit into the category of bundling of securities, while others will be more transaction oriented. However, in a few areas, even functional definitions will not apply.

Nevertheless, the function-based definitions should be retained for three important reasons. First, the traditional breakdown of the industry (commercial and investment banking, insurance, mutual funds) is no longer satisfactory. Second, the functional breakdown may be important for future academic research, because some areas, such as encryption for electronic transactions, clearly have very large engineering components. Other areas, such as the pricing and packaging of new types of securities, have more of a finance/economics/financial engineering component. Third, government regulation will be promulgated according to where the boundaries are drawn.

INNOVATION SYSTEM

Changes in the past 20 years have fundamentally changed the markets, the way customers interact with service providers, and the range of products in the financial services industry. These changes have taken place in a relatively short period of time and with no organized innovation process or formal organizational units devoted to research like the ones in most science- and engineering-based industries. Historically, however, many science- and engineering-based industries did not have structured innovation systems in the early stages of their development. If the history of these industries is any guide, developments of the last several decades in financial services, in which technology has become critical to success in the industry, could represent the beginning of a shift to an organized research base. In the meantime, the absence of an organized research base has not hindered rapid innovation.

Industry Drivers

There are three primary drivers of innovation in financial services: market forces; tax and regulatory policy; and technology, particularly data processing, communications, and software. Often, these drivers interact. For example, the spread of the Bloomberg and other reporting services were made possible by technology, but were driven by market demand, which was reinforced because the availability of this information made the markets work better.

A second pathway for innovation is regulatory reform, usually triggered by a large loss or failure. Pension plans, for example, are highly regulated to avoid repetition of earlier mismanagement of pension assets. The Employee Retirement Income Security Act (ERISA) requires that actuaries conduct independent evaluations of all defined pension plans each year and estimate the plan's surplus or deficit, based on a careful analysis of assets and liabilities over a long time horizon. Meeting this requirement provides employment for tens of thousands of actuaries.

Insurance companies must perform similar studies. The insurance industry is even more complicated because all 50 states have insurance commissions with separate requirements. Insurance companies must report their statutory surplus/deficits (STATs), their surplus/deficit according to generally accepted accounting principles (GAAPs), and sometimes their economic surplus/deficits. Tens of thousands of accountants, tax advisors, and related professionals are employed in these studies.

Periodically, new regulatory requirements are added. According to the Financial Accounting Standards Board (FASB) and its standards, such as FASB 87, pension plans in the United States must use a close-to-market discount factor when computing the market value of liabilities, rather than a fixed or arbitrary discount rate. This change is consistent with market-driven reporting but can cause large swings in surplus/deficits because of changes in long-term interest rates. As a consequence, pension plans must be more careful in allocating their assets and may purchase new financial products, such as complex options, to help them achieve their goals. As a rule, when regulators impose new burdens on investors, the financial services industry responds by offering new products and services.

Another example of regulation-driven innovation is the ever-changing tax code. Tax-advantaged investing is possible through a myriad of options—traditional individual retirement accounts (IRAs), Roth IRAs, Keogh plans, state college tuition plans, variable annuities, other insurance-related investment vehicles, and many others. University researchers sometimes play a key role in these innovations. The U.S. Treasury Department's Office of Tax Analysis (OTA) routinely hires university faculty members for one- to three-year assignments. During a major study on tax reform, for example, David Bradford (of Princeton's Woodrow Wilson School) was head of OTA. Faculty advisors are also shaping the current debate on social security reform. New financial products will certainly emerge as the government provides new avenues for deferring taxes or shifting ordinary income into capital gains.

Although regulatory constraints can accelerate innovation by creating incentives to meet them (or work around them), regulations per se are not the principal drivers of innovation. Competitive pressures and innovations enabled by advances in computing and communications technology and the modeling and simulation of financial markets are much stronger drivers.

Institutional Drivers

Academic research has played an important role in innovation in the financial services industry, but a number of other players and factors have also contributed significantly. The players are (in no particular order) financial services companies, universities/academic institutions, government agencies (oversight regulations and taxation), software developers, hardware manufacturers, spin-off companies, and consulting companies.

Financial Institutions

Innovation in investment banking has generally been initiated at major companies and has then been diffused to smaller companies and individuals. Large banks with large budgets for new or improved technology are able to support both internal and external research. The results then trickle down to the overall market and smaller competitors. Much of the research is performed at various levels within the banks. Some of the work is undertaken in response to a central corporate directive, but most of it is done in the operating units with no particular coordination. In general, the most productive R&D in the industry has taken place in a few pockets in larger firms and has been loosely organized. The disciplined ethos of R&D that can be found in high-technology manufacturing industries is totally lacking in financial services.

Consultants and Consulting Firms

Consultants and consulting firms have contributed significantly to innovation in financial services, sometimes based on academic research. More and more of these firms are developing internal quantitative-research capabilities. Management consulting firms, for instance, have developed the following innovative frameworks for measuring a company's economic performance:

- activity-based costing (a refinement of standard accounting practice that allocates costs to specific business activities)
- economic value added (EVA®) (an approach used by Stern Stewart & Co. to measure the economic profit of an enterprise by calculating net-operating profit minus an appropriate charge for the opportunity cost of all capital invested in the enterprise)
- cash-flow return on investment (CFROI) (an approach used by Boston Consulting Group to compare a firm's cash flows with the inflation-adjusted capital used to produce them)
- balanced scorecards (developed by David Norton, a consultant, and Robert Kaplan, a professor at Harvard Business School) to combine financial scores with measures of less tangible assets, such as customer satisfaction and loyalty, and a firm's ability to nurture the skills of its employees)

Government

Government has played a limited role in funding and using research in financial services. Government has, however, funded the development of some crucial infrastructure technologies, such as check imaging. In addition, government responsibility for the safety and soundness of the financial system has been a driver of economic and mathematical research in risk measurement and risk management. The resulting improvements in risk measurement and modeling technologies played a significant role in the decision to repeal the Glass-Steagall Act of 1933. In the 1970s, government-funded research developed models for analyzing tax policy, which provided systems for analyzing the impact of changes in the tax code on different groups in the United States. Today, these models are used by every government agency.

Although government funds only a small amount of academic research relevant to financial services, it does support research in economics and mathematics ($89 million and $229 million respectively in federal funding in 2000). Total funding for academic research from federal and nonfederal sources was $255 million in economics and $341 million in mathematics (NSF, 2003).

Software Developers and Hardware Manufacturers

The financial services industry relies heavily on technology developed in other industries, especially information technology, computers, communications equipment, and software. In a compilation of R&D spending on information technology in 500 end-user organizations, banking and financial services had the largest budget (9 percent) as a share of projected revenues and were estimated to have spent the largest percentage (0.45 percent) of projected sales revenues on information technologies R&D (NRC, 2000).

Investment in information technology has resulted in software development that has enabled much of the innovation in financial services. For example, Algorithmics is a growing software firm specializing in risk measurement and management software. Dr. Ron Dembo, the founder and CEO of Algorithmics, was formerly on the faculties of several universities including Yale, MIT, and the University of Toronto.

Industry Consortia and Cooperation

Financial services have traditionally worked together in moving funds, setting standards, facilitating cooperation, and sharing risk. However, sharing information in financial services is not easy. Because of competitive secrecy and the lack of protections for intellectual property, cooperation has not been focused on innovation. Because of competitive secrecy, it is difficult to get a sense of how much interesting and valuable research is being performed within the industry.

The threat of competition from companies that have access to new innovative technologies has encouraged some attempts at cooperation in precompetitive research. The Financial Services Technology Consortia (FSTC) was established in 1993 to undertake precompetitive, collaborative R&D in electronic commerce, cryptography, education, and other areas. This initiative marked a first attempt at collaboration in R&D by the industry and mirrored the same trend in technology industries. The American Bankers Association followed with its own technology initiative, the Banking Industry Technology Secretariat (BITS), a division of the Bankers Roundtable. Another example is the Smart Card Alliance, in which more than 185 companies participate. The Smart Card Alliance was formed by the

2001 merger of the Smart Card Forum, comprising primarily financial services firms, and the Smart Card Industry Association, whose members include a variety of industries and government agencies.

In 1999, a risk management initiative was launched by 12 of the world's leading banks, 11 of which had been part of the consortium that provided Long-Term Capital Management with $3.5 billion following its near collapse in the fall of 1998. The risk-management group, called the Counterparty Risk Management Policy Group, included Barclays Bank PLC, Bear Stearns & Company, Chase Manhattan Corporation, Citigroup, Inc., Credit Suisse First Boston, Deutsche Bank, Lehman Brothers, Inc., Merrill Lynch and Company, Inc., Morgan Stanley Dean Witter, and Warburg Dillon Read, a unit of UBS AG. The stated purpose of the group is to enhance existing risk-management practices and to compile information that could be reported to "regulators and supervisors, all with a view to enhancing the discipline, efficiency and liquidity of financial markets." (Bloomberg Information, 1999). The group issued a report in June 1999 that included more than 20 recommendations to improve risk management (Counterparty Risk Management Policy Group, 1999).[3]

Intellectual Property

Historically, protecting intellectual property was not an issue in financial services because there was little opportunity to create it and, therefore, little need to protect it. When new financial instruments were created, they often had very short half-lives, so the emphasis was on being first to market with innovations and capturing market share before competing products were introduced. Trading systems and databases are considered highly proprietary and are regarded as trade secrets. This industry posture vis-à-vis intellectual property (e.g., the need for secrecy, tight controls) has probably been an impediment to collaborations with academic researchers.

The potential to create competitive advantage through patenting was successfully demonstrated when, in 1980, Merrill Lynch filed for and was issued U.S. Patent No. 4,376,978. The patent relates to the Merrill Lynch Cash Management Account program (CMA) and, specifically, the data processing methodology and programs for effecting the CMA. The CMA consolidated checking, savings, investments, and borrowing into an all-in-one account, which is invested in a money market fund. The patent allowed Merrill Lynch to exclusively offer the CMA; later the company gained substantial revenues by licensing the same technology to other securities firms.

Recently, banks and other traditional and aspiring financial institutions have become interested in patenting in the area of financial services, due in part to the 1998 "State Street" decision by the Patent Court, which broadened patent protection to include "business methods," including financial services. Although much of the patent activity is by technology companies outside financial services, the top two assignees for patents are Merrill Lynch and Citicorp (Lerner, 2000). Today, many new patents are being issued, some involving systems and technologically based financial innovations, but only one of the top five companies with patents in the area of electronic commerce is a financial services company. It remains to be seen if these patents are enforceable.

Human Capital

Traditionally, bankers were regarded as predominantly "relationship people" who were not mathematically oriented or analytical. Bankers today, however, use financial applications based on the Black/Scholes model for pricing financial options (Black and Scholes, 1973). Because of regulatory changes, technological innovations, globalization, and other factors, bankers must now understand and react to events in financial markets. This has resulted in a significant increase in the demand for scientists and engineers. For example, in 1997 when the Federal Reserve imposed new capital/reserve requirements based on a value-at-risk system, it created a demand for people who could calculate these requirements on an integrated basis. At the same time, science and engineering graduates with new knowledge and new skills enabled the Federal Reserve to develop these new reserve-requirement formulas.

Scientists and engineers are generally not hired to perform research in financial services companies. They are usually scattered throughout the organization, including the small business units, where they are often indistinguishable from traders and other personnel.

CONTRIBUTIONS OF ACADEMIC RESEARCH

The links between academic research and the financial services industry are not nearly as well established as they are in many other industries. Nevertheless, academic research has made very important contributions to the industry. There is a tendency to overstate the importance of breakthrough innovations in financial services and to understate the importance of a long list of small but significant incremental innovations. The aggregate consequences of academic research for the financial services sector cannot be quantified, but examples can be cited to illustrate their impact.

Academic research on financial engineering[4] has been especially useful to the industry. Because underlying theory and actual practices are well aligned in this area, a steady stream of academic contributions has flowed to the industry (Heath et al., 1992). Related academic research has also contributed to other kinds of modeling (e.g., volatility skew), binomial methods and their extension, and partial differential equation methods. Monte Carlo methods, originally developed in the national laboratory system for weapon design, have been advanced through excellent cross-fertilization between industrial and academic research.

At the same time, research in the traditional disciplines of finance and economics has also made substantial contributions to financial services. Examples include conceptual breakthroughs such as portfolio diversification and "two-fund separation" theory, which helped lay the foundation for modern index funds; the development of optimal capital-structure theory; contributions from behavioral finance on decision making under uncertainty; as well as such widely used tools in econometric and economic modeling as the generalized method of moments estimation. The financial services industry can be widely defined to include just about all business applications dealing with financial information. Our focus on the narrow area of financial engineering illustrates the relationship of academic research and subsequent applications in industry. Additional linkages could be cited, depending on the definition of financial economics and financial engineering.

The nature of academic contributions to the financial services sector is likely to change in the future. As the industry evolves, new areas of academic research are likely to become useful. For example, academic research in consumer behavior that sheds light on how individuals make decisions regarding financial products and services has become extremely important. As large amounts of data derived from networks, databases, and transaction systems become available, the industry is likely to use more advanced, nonlinear mathematical analyses and other methods to predict consumer behavior.

Areas and Types of Contributions

Major contributions of academic research to financial services can be traced back to five areas: conceptual breakthroughs; financial products and tools; consumer research; research on legal, regulatory, and institutional issues; and research on industry infrastructure.

Conceptual Breakthroughs

Fundamental research in portfolio theory, linear programming, derivative pricing theory, and prospect theory laid the foundation for entirely new families of financial products and services. At least nine Nobel prizes in economics have been awarded to researchers associated with universities for major conceptual breakthroughs important to financial services:

- Daniel Kahneman (Princeton) in 2002 for integrating insights from psychological research into economic science, especially concerning human judgment and decision making under uncertainty. Along with Amos Tversky (Stanford), this work created the field of behavioral finance.
- Myron Scholes (Stanford) and Robert Merton (Harvard) (with Fisher Black of Goldman Sachs) in 1997 for a new method of determining the

value of derivative securities. The breakthrough idea was that derivatives can be priced without estimating the future direction of the underlying security. This concept spawned a new industry—the development, pricing, and trading of derivative securities.

- John Harsanyi (University of California, Berkeley), John Nash (Princeton), and Richard Selten (Rheinische Friedrich-Wilhelms-Universität Bonn) in 1994 for pioneering analysis of equilibria in the theory of non-cooperative games. Non-cooperative game theory excludes binding agreements while making predictions about the outcome of strategic interactions. It has become a standard tool in the analysis of oligopoly and the study of competition between firms in the theory of industrial organization. But the concept has also been used in macroeconomic theory (e.g., for economic policy, environmental and resource economics, foreign trade theory, the economics of information) to make complex strategic interactions more understandable.
- Harry Markowitz, Merton Miller (University of Chicago), and William Sharpe (Stanford) in 1990 for their work on portfolio theory using quadratic programs to explore diversification strategies to reduce risks. Concepts based on efficient frontiers and mean-variance are used to find a mix of assets that simultaneously maximizes the investor's return and minimizes risks. This theory provides the basis for a large number of financial planning packages.
- Merton Miller (University of Chicago) in 1990 and Franco Modigliani (MIT) in 1985 for their seminal contributions to the development of optimal capital-structure theory, specifically for positing the theory for the capital-market relationship between the capital-asset structure and dividend policy of production firms, and firms' market value and costs of capital. This work provides a basis for structuring corporations in highly competitive markets.
- James Tobin (Yale) in 1981 for his work on portfolio diversification and "two-fund separation theory," which along with the Capital Asset Pricing Model, developed by William Sharpe, provided the conceptual foundations for index funds.
- Tjalling Koopmans (Yale) and Leonid Kantorovich (Moscow Academy of Sciences) in 1975 for optimum resource allocation theory, which provides a practical approach for managing large, complex organizations and is used by many financial companies for managing assets and liabilities.[5]
- Kenneth Arrow (Harvard/Stanford) in 1972 for foundational work on economic equilibrium and state-spaces models of uncertainty.
- Paul Samuelson (MIT) in 1970 for his role in bringing quantitative reasoning to economics and for his work on the implications of optimization for understanding the qualitative behavior of models.

Financial Products and Tools

Academics have been involved in applied research to develop specific products in financial engineering and optimization, such as pricing models and portfolio methods. Although relatively few financial services firms have specialized personnel capable of incorporating the results of academic research, a growing number of firms are convinced of their value for real-world business. Some of the most important products and tools are listed below:

- foundations of option pricing (Black and Scholes, 1973; Merton, 1973)
- numerical methods of pricing and hedging, including discrete approximations of diffusion processes, partial differential equation methods, and Monte Carlo survey methods
- arbitrage restrictions on the term structure of interest rates, especially the standard interest rate model developed by David Heath, Robert Jarrow, and Andrew Morton (1992) at Cornell University, and earlier work by Oldrich Vasicek (1977), Thomas Ho and Sang-Bin Lee (1986), and others
- formulation and solution to problems of portfolio optimization begun by Harry Markowitz when he was a doctoral student at University of Chicago and continued by many others, such as Michael Dempster at Cambridge, John Mulvey at Princeton, Stavros Zenios at the University of Cyprus, and William Ziemba at the University of British Columbia
- methods of analyzing financial time series, especially models that incorporate autocorrelation and varying volatility (e.g., autoregressive conditional heteroskedasticity [ARCH] models, introduced by Robert Engle, New York University, in 1982, and their variants)
- generalized method of moments estimation tools that are widely applied to a large number of econometric and economic models (although the method of moments has been known for more than 100 years, L.P. Hansen [University of Chicago] extended the concepts in what became known as the generalized method of moments) (Hansen, 2001)
- use of mechanism design theory, which studies the design of institutions or mechanisms implementing collective decisions under different circumstances, for contractual design
- incorporation of credit risk into pricing models (e.g., Jarrow and Turnbull, 1995; Duffie and Singleton, 1997)

Consumer Research

Consumer research began about 25 years ago and is now very common. Academic research has contributed to an understanding of large data sets, consumer behavior, and so on. The closest academic counterpart to industry research on consumer financial behavior is in the social sciences, which tend to focus on aggregate human behavior, but rarely cover financial behavior. Although marketing is taught as an academic subject, there are few, if any, programs on consumer (credit) finance. The University of Michigan Panel Study of Income Dynamics covers some financial behavior, and the Credit Research Institute also does some work in this area.[6] But little academic research focuses directly on how financial institutions can influence customer behavior (business strategies), how technical tools can be used, or the best strategies for affecting consumer portfolio behavior.

Academic research has contributed to the development of data and technical tools for personal financial management but has had little influence on teaching people how to analyze data or use the tools effectively. Unlike investment banking, there seems to be little underlying social theory in this area. The trend in consumer financial institutions is to hire scientists (e.g., physicists and mathematicians) and other highly trained technical people who often have little experience in dealing with consumer behavior and the fuzziness of action/reaction patterns. Research in this area could be very helpful to policy makers, who know very little about this area of social behavior or have ideas based on outdated studies. Research could improve their understanding of which segments of the population are able to manage their own financial portfolios to maintain adequate income through retirement, which segments require public management through Social Security and Medicare, how public programs can be made more efficient, and how changes in public programs can be made.

In 1996, the Employee Benefit Research Institute (EBRI) arranged a review of public-sector and private-sector dialogues to call attention to the lack of good research on retirement issues. EBRI and the Investment Counselors Institute managed to create a shared database on 401(k) participants to inform more effective policies.

Legal, Regulatory, and Institutional Issues

Academic researchers have been instrumental in analyzing legal and regulatory constraints on the industry and in reforming the regulatory environment. Academia-developed models of asset and arbitrage pricing and risk management have been extremely influential. Rules for trading activity are based on value-at-risk models. Because of the complexity of many global markets, such as energy, securities, and foreign exchange, regulators and economic policy makers now depend on models as a basis for decisions, which provide an objective basis for making decisions in real-world markets that are too diverse or take too long to provide feedback. In these cases, the models do not simply represent reality; they become reality.

Infrastructure of the Industry

Academic research has contributed to the infrastructure of the modern financial services sector, particularly in the area of encryption technology and networking. The technological infrastructure of the financial services industry is probably the most extensive and complex of any service sector in the U.S. economy. Academic contributions to networking technology have had a fundamental impact, enabling electronic payment systems, automated teller machines, and the growth of credit and debit cards in the 1970s and the emergence of electronic commerce via the Internet in the 1990s. (For more about networking technology, see discussion in Chapter 2.)

TRANSFER MECHANISMS

A pattern has developed for transferring the results of academic research to the financial services industry. First, academic researchers publish a series of papers on a topic in the field of financial economics. The papers set the stage for a few innovative firms to test products based on the idea; faculty members are often involved as consultants to these firms. In some cases, junior and senior faculty resign from academia to work on these projects full time. If the product proves to be effective, the financial industry invests in further development. At this point, many firms attempt to duplicate the product or service. Controversies about the protection of intellectual property via trademarks, copywriting, trade secrets, and patents are addressed in the courts as they arise.

An excellent example of this pattern involves index funds. During the 1970s and 1980s, there was considerable debate among academics about the efficiency of financial markets and the ability of portfolio managers to succeed over long periods of time. Some argued that a market fund could outperform the majority of active portfolio managers. Wells Fargo and several other innovative banks set up mutual funds with the sole aim of mimicking the return of the S&P 500 stock index. (Often the banks would purchase the stock in all 500 companies). At the urging of Professor Burt Malkiel at Princeton University, the Vanguard Company established the first index fund for individual investors. Since then, index funds have become enormously popular because of their low cost structure, their availability, and their historical performance. For example, Vanguard's Total Stock Market Index Fund had a net asset value of $15.8 billion on January 31, 2002. Fidelity's S&P 500 index fund had a net asset value of $8.6 billion at the end of January 2002. These funds are directly attributable to the early research of academics.

A significant portion of current activities in the financial services sector would not be possible without fundamental mathematical tools that were developed or adapted to financial problems by academics. Beginning with the Black-Scholes option-pricing formula now applied in a myriad of ways, these mathematical techniques enable the industry to price an almost unlimited variety of financial instruments. Markets as diverse as options, futures, other derivatives, securitization, and reinsurance could not exist in their current forms without these tools. Today, these techniques are widely known and understood, and new

TABLE 6-1 Innovations in Financial Services Resulting from Academic Research

New financial instruments/securities/products
- index funds
- financial derivatives
- securitization (risk neutral valuations)
- credit derivatives (emerging)/credit management (modification of models developed for science and engineering)

Analytical and Modeling Tools
- option pricing (black-scholes and successors)
- portfolio theory
- market microstructures

Financial Information and Research Tools
- online information services (mixed academic-industry heritage)
- online investment services
- off-the-shelf risk-management software
- risk metrics/credit metrics

Transactions
- cryptography
- electronic commerce/world wide web (mixed academic-industry heritage)

Consumer Research
- scoring models
- panel studies

products and applications are being developed without further interaction with the academic sector (although the work is done by individuals trained in academia). Table 6-1 shows innovations in financial services that were generated at least partly by academic research.

Flow of Ideas

Advice and ideas from academia to the private sector continue to flow. Commercial and investment banks rely heavily on academic economists, who have a substantial influence on risk management because they are instrumental in the development of techniques and modeling tools and have insights into the macroeconomic developments that are inputs to the models. A mutually reinforcing dynamic is at work among the diffusion of human capital into the financial services firms, the development of increasingly powerful computer hardware, and the insights of academics into financial theory. For example, the combined capabilities of technology and academic research were involved in the development of the structure of large-scale auctions, such as the auctioning of the radio-frequency spectrum.

Flow of Human Capital

The flow of human capital into financial services firms has been crucial to the evolution of the industry, which draws on people from a wide variety of disciplines who have mathematical and modeling skills. It is not clear whether this diversity is a plus or a minus for innovation. On the one hand, considerable "on-the-job training" is necessary before a physicist or nuclear engineer can apply his or her modeling skills to the analysis of financial problems. On the other hand, people with diverse backgrounds who have been encouraged to think creatively by graduate research may be a source of creativity and may come up with new approaches to problems.

It is probably inevitable that training for employment in the financial services industry will gradually become more traditional. As finance and financial engineering programs become more widespread and more firmly established, they will probably provide a steady flow of people with postgraduate degrees targeted specifically to the financial sector.

In general, although academic research is stimulating to some, it can seem esoteric and confusing to others. Although much of the academic literature is all but incomprehensible to nonacademics, the dissemination of research results is essential for continued innovation. So-called "quants" (scientific quantitative people) working in the derivatives industry are most likely to try to keep abreast of academic research by reading preprints and attending conferences.

To bridge the academia-industry gap and to meet the need for more avenues of transfer, some have argued that applied research should be done in individual companies (loosely linked to a university) or through university think tanks.

Impact of Academic Research

Links between academia and the financial services industry are less formal, less structured, and more amorphous than the links in science-based industries. This may be because innovation in financial services is inherently less amenable to structured, scientific investigation. It could also be that this is a natural stage in the evolution of the industry and that, as academic disciplines such as financial engineering evolve, the structure of the innovation system will evolve with them. If so, the current transitional relationships between academia and the financial services sector will be replaced by a more structured innovation system in the future.

The financial services industry can be divided into two fundamental areas: (1) securitization, or the packaging and bundling of financial products or services for consumers or businesses (security bundling); and (2) the facilitation of trades, or helping to implement transactions (transactional). Academic research has had a more direct impact on securitization (i.e., with development of novel financial products and services). Clearly, financial instruments, investment management, and decision support tools created in academia or through public-private-sector partnerships have been critical to the success of the financial services industry.

Quantitative measures of the impact of academic research are difficult to determine. The entire industry is characterized by phenomenal increases in the amount of quantitative data concerning transaction volumes, flow rates, asset bases, customer activity, product ranges, and other activities, but assigning specific increases to academic research is difficult. Certainly the number of new securities introduced each year based on academic research is large and growing, and the number of institutions and people who use these products for managing their financial affairs can be estimated. For example, the dramatic increase in option trading in the past decade is largely attributable to the efforts of academic researchers. Another measure is the increase in the number of people who are working on the development of financial products and in related areas. Finally, it may be possible to measure improvements in economic efficiency. Although the task is complicated by the heterogeneous nature of the industry, an informed, intuitive review suggests that much of the recent growth in the industry can be traced to academic research.

Some have argued that academic research can also be considered to have had a negative impact on the industry. Some have argued that a number of recent financial disasters, such as the near demise in 1998 of Long Term Capital Management, a very large, global hedge fund that was heavily influenced by academic research, can be attributed to the results of research in financial engineering. This experience demonstrated that the pricing formulas used in many financial models do not account for periods of economic turmoil and alerted the financial community to these pitfalls and limitations.

OPPORTUNITIES FOR FUTURE ACADEMIC RESEARCH

A broad range of academic research in engineering, natural sciences, economics, mathematics, social sciences, and public policy will continue to contribute to the success of the financial services industry:

- identification of rogue traders
- allocations of capital to various activities
- the dynamics of markets (stochastic models) and market microstructures
- evaluation of mergers of financial services providers
- issues of globalization, such as diverse regulation, capital flight driving countries to near bankruptcy, tax havens, and money laundering
- investigations of legal, sociological, and technological issues related to privacy, trust, security, contract law, etc.
- assessments of the impact of technology
- risk audits
- ethical dimensions of finance

FINDINGS AND RECOMMENDATIONS

The results of academic research in economics, information systems, and other areas have had a substantial, direct impact on the structure and performance of the financial services industry in the past 20 years, especially in the areas of risk management and new financial instruments, such as derivatives. The panel has attempted to distinguish between the direct contributions of research (e.g., modeling and risk management) and indirect contributions (e.g., contributions of people educated and trained at universities and research results embodied in software and hardware used by the industry). Indirect contributions, such as research on cryptographic algorithms that now play a critical role in financial services, have been significant. Both fundamental and applied research have affected the industry. In some areas, such as the pricing of securities, academic research has played a major role. In other domains, such as the identification of arbitrage opportunities, economic incentives are more than adequate to keep researchers busy. These are important distinctions, because funding for academic research should be focused on the exploration of basic concepts that will not be undertaken under the current incentive system. A natural tendency has been to look for specific engineering or applied research influences in financial services. However, a considerable body of academic research, including fundamental research in economics and finance, has been applicable to financial services. Connecting the academic research base with the financial services industry remains an ongoing challenge as well as an important opportunity.

Although it is difficult to predict the future significance of academic research, the increasing innovation and complexity of financial services will probably prove to be a rich and fertile area for future academic research. For example, the emergence of global consumer markets for financial services could create research opportunities in consumer research and behavioral sciences. Overall, it is difficult to compare the importance of academic research to the importance of other sources of new knowledge and innovation. It is evident, however, that the contributions of academic research have been fundamental and consequential.

Finding 6-1. Technology derived from academic research in economics, information systems, engineering, and other areas has been fundamental to innovation in financial services in the last 20 years.

Technology can be defined broadly to include pricing securities and risk-management systems. In the past, academic research relevant to financial services was concentrated on fundamental concepts in finance, economics, and technology. Now, academic research on applications from engineering disciplines and the physical sciences has become more important, and the number of joint university-industry activities in these areas has increased significantly.

Opportunities for the application of engineering methods and processes in the financial services industry are expanding rapidly. With the advent of mass-scale consumer marketing involving tens of millions of transactions, financial services are becoming similar in some ways to manufacturing and process industries. The advent of the Internet and its potential transformation of financial services has underscored the role of engineers and technology to the future of financial services. At the same time, engineers can also learn from engaging with the financial services industry, particularly in the areas of stochastic and model-based approaches to problem definition and solution.

Recommendation 6-1. The National Academy of Engineering together with other agencies or private foundations should examine how engineering methods and processes practiced in advanced manufacturing industries can be applied in financial services.

Finding 6-2. Because financial services firms traditionally have had no formal research function or culture (in contrast to science-based industries), no formal organizational structure has been established for technical personnel to keep track of academic research or to communicate industry research needs to academia. At the same time, academia has also failed to establish an organized way of mapping the needs of the industry and relating them to academic research.

Recommendation 6-2. The management of financial services companies should become more familiar with the ways academic research affects their businesses. Managers should review the processes they have (or need) to ensure that they can take advantage of the results of academic research. The creation of an engineering research center for financial services, similar to the research centers supported by the National Science Foundation for engineering, could provide a nexus for industry-academic activities.

Finding 6-3. A convergence of events, including deregulation, globalization, increasing competition, and the information revolution, has made financial services more technically intense and accelerated innovation in the industry. The panel expects innovation to remain very rapid, driven mainly by developments in engineering and technology.

Recommendation 6-3. Professional societies, policy research organizations, and federal agencies associated with science, engineering, and technology, including the National Academy of Engineering and the National Science Foundation, must recognize the importance of research and technology to financial services and include financial services in their mainstream activities. Accordingly, federal funding for long-term, potentially high-impact research relevant to financial services, which is currently very limited, should be expanded.

Finding 6-4. Regulation has been a major factor shaping the financial services industry, both as a boundary condition and as a stimulus to innovation. The impact of academic research on regulation has been relatively small historically (and usually in response to a crisis); however, the impact of academic research on regulation, particularly on risk management, is growing. In general, the regulatory community lags behind the industry proper in the application of new knowledge and tools in the area of risk management.

Recommendation 6-4. Regulators should be encouraged to support more academic research in their areas of concern and to become more knowledgeable in modern risk-management methods.

Finding 6-5. The responsibility for financial decisions is being shifted to individual consumers of financial products, and individual responsibility for financing retirement is becoming increasingly likely. The current administration's proposal for partially privatizing Social Security is a move in this direction. The knowledge and tool base available to help individuals make good financial decisions is in its infancy. Most marketing research is descriptive and is performed for the benefit of financial service providers; very little normative research is aimed at helping individual consumers make better financial decisions.

Recommendation 6-5. The National Science Foundation and other federal agencies should fund normative research to help consumers make better financial decisions to complement anticipated legislative action that would shift financial responsibilities to individuals.

Finding 6-6. Individuals play an important role in transferring the results of research from academia to the financial services industry. As the field of finance becomes more quantitative and analytical, the training of engineers and financial analysts is becoming increasingly similar. The migration of academically trained people to the industry has apparently had a significant impact on successful innovation. Unlike more science-based industries, financial services companies have only recently begun to hire technically trained people who can interact profitably with academic researchers and translate industry needs to academia and academic research results to industry.

Today, academia provides much of the human capital for the financial services industry. Business schools, for instance, train large numbers of practitioners for the industry, including many who advance to senior management levels. Engineering departments train many of the analysts, but there is very little support or guidance for this training from the financial industry. Although the industry benefits greatly for very little investment, the benefits could be even greater if the industry provided more support and was more closely involved.

Recommendation 6-6. The financial services industry should encourage research exchanges, sabbaticals in industry, industry practitioners teaching in universities, and other industry-academic interactions. The industry would benefit greatly from the establishment of mechanisms to facilitate the flow of people between academic research and the financial services industry. Mechanisms could include more targeted curricula and programs in computational finance, financial engineering, and other areas; revised incentive and reward systems; and protections of intellectual property.

Finding 6-7. Historically, intellectual property in financial services has not enjoyed the same level of protection as intellectual property in science-based industries. Instead, the industry has focused on keeping trade secrets and speed to market. Although patenting has increased greatly, it is too soon to tell if this will have a positive or negative effect on innovation and growth. Because technology is essential to many innovations, more patents are sure to be granted.

Recommendation 6-7. The panel calls upon the U.S. Patent Office, Federal Reserve, U.S. Department of the Treasury, National Science Foundation, and other affected federal agencies to support research on the impact of changes in the treatment of intellectual property in the financial services industry, specifically the impact of the recent flurry of patent activity.

Finding 6-8. The relationship between innovation and economic performance is poorly understood. This problem is not specific to the financial services industry.

Recommendation 6-8. Given the sheer size and importance of the financial services sector to the nation's economy, measures of innovation and of the relationship between innovation and economic performance should be developed for the financial services industry. Although defining the end product of financial services will be difficult, focused research on this problem could reveal suitable measurements.

Finding 6-9. The deregulation of markets has created many opportunities for businesses and individuals but has also increased risks. For a variety of reasons, including technological advances, esoteric financial instruments and structures, globalization, the speed and magnitude of money/asset movement, and mass communications, risk management is more important for financial services now than it was in the past. Risks, ranging from operational risks to credit and currency risks, are more numerous and less understood than ever before. Each risk has the potential of endangering not only financial institutions but also sovereign nations and, ultimately, the global economic system.

Recommendation 6-9. Management in financial services should review how leading engineering-intensive industries have integrated risk management and quality improvement into all aspects of the design, production, and delivery of their products and should apply the principles to financial services. The National Academy of Engineering should convene a meeting to develop a suitable research agenda.

Finding 6-10. A research gap has become apparent between basic research and applied research, particularly in the areas of financial models and market reality, and the financial infrastructure and its behavior under various conditions. To bridge the gap between theory and practice in experimental economics, economic theories should be tested in real markets and the structures analyzed before new (potentially destabilizing) products are introduced. Although regulators have tried to anticipate problems by using models and other tests, their efforts have been hampered by inadequate data and a lack of established processes. In general, the industry does not have a solid theoretical foundation to mount an institutional response to problems. The lack of data is largely the result of companies being unwilling to share data on credit risk, operational risk, corporate defaults, and other aspects of their businesses.

Recommendation 6-10a. To make data more widely available to academia, data from competing financial institutions should be collected by third parties and "sanitized." One simple idea would be for the industry to fund a web service that would help individual researchers find the sources of data in financial services. Users of this data could include off-Wall Street firms and, of course, individuals, who are increasingly being urged to manage their own financial futures.

Recommendation 6-10b. Financial services companies should establish consortia/cooperative research to define and build a technological infrastructure and knowledge base for various sectors of the financial services industry. Cooperative activities could be based on insights from Bankers Roundtable Information Technology Secretariat (BITS), Financial Services Technology Consortium (FSTC), Smart Card Forum, and Counterparty Risk Management Policy Group.

NOTES

[1]AT&T sold its Universal Card business to Citicorp in December 1997.

[2]It is not at all clear that available financial advice, tools, and data are sufficient for individuals to take control of their financial future. Should investment management by professionals, always controversial in terms of performance, be replaced by investment management by individual consumers? Can an individual consumer exposed to information from many sources understand "real-time risk" as financial and economic events unfold almost in real time on television or the Internet? In the wake of stock market declines in 2000 and 2001 and the Enron debacle in 2002, the answers to these questions have assumed real policy implications.

[3]In June 1999, representatives of the Counterparty Risk Management Policy Group presented policy recommendations to the Subcommittee on Capital Markets, Securities and Government Sponsored Enterprises of the House Committee on Banking and Financial Services.

[4]Financial engineering, the quantitative analysis of financial markets, complex securities, and risk management, using mathematical, statistical, and computational models, is now included in the curriculum at many universities.

[5]George Dantzig's development of the simplex method should also be noted.

[6]The Panel Study of Income Dynamics (PSID) at the University of Michigan is a longitudinal survey of a representative sample of individuals and their families. The study has been ongoing since 1968. The data were collected annually through 1997, and biennially since 1999. The data files contain the full span of information collected over the course of the study. PSID data can be used for cross-sectional, longitudinal, and intergenerational analysis and for studies of individuals and families.

REFERENCES

Black, F., and M. Scholes. 1973. The pricing of options and corporate liabilities. Journal of Political Economy 81(3): 637–654.

Bloomberg Information. 1999. Securities Firms and Banks Form Risk Management Group. January 7. Available online at : *http://www.bloomberg.com/.*

Counterparty Risk Management Policy Group. 1999. Improving Counterparty Risk Management Practices. New York: Counterparty Risk Management Policy Group.

Duffie, D., and K. Singleton. 1997. An economic model of the term structure of interest rate swap yields. Journal of Finance 52(4): 1287–1321.

Federal Reserve. 2002. Assets and Liabilities of Commercial Banks in the United States. Federal Reserve Statistical Release. Washington, D.C.: The Federal Reserve Board.

Hansen, L.P. 2001. Method of Moments. In International Encyclopedia of the Social and Behavior Sciences, N.J. Smelser and P.B. Bates, eds. Amsterdam and New York.: Elsevier.

Heath, D., R. Jarrow, and A. Morton. 1992. Bond pricing and term structure of interest rates: a new methodology for contingent claims valuation. Econometrica 60: 77–105.

Ho, T.S.Y., and S-B. Lee. 1986. Term structure movements and pricing interest rate contingent claims. Journal of Finance 41: 1011–1029.

Jarrow, R., and S. Turnbull. 1995. Derivative Securities. Cincinnati, Ohio: South-Western College Publishing.

Lerner, J. 2000. Where Does State Street Lead? A First Look at Finance Patents, 1971–2000. Working Paper No. w7918. Cambridge, Mass.: National Bureau of Economic Research.

Merton, R. 1973. Theory of rational option pricing. Bell Journal of Economics and Management Science 41: 141–183.

NRC (National Research Council). 2000. Making IT Better: Expanding Information Technology Research to Meet Society's Need. Washington, D.C.: National Academies Press.

NSF (National Science Foundation). 2003. Academic Research and Development Expenditures: Fiscal Year 2001. NSF 03-316. Arlington, Va.: National Science Foundation.

U.S. Bureau of the Census. 2002. Statistical Abstract of the United States–2002. Washington, D.C.: U.S. Government Printing Office. Available online at: *http://www.census.gov/prod/www/statistical-abstract-02.html.* [June 24, 2003]

Vasicek, O. 1977. An equilibrium characterization of the term structure. Journal of Financial Economics 5: 177–188.

ADDENDUM

Questionnaire

The following questionnaire was sent to selected individuals from various parts of the financial services industry, some of whom attended the October 1998 workshop. Included among the questionnaire respondents were senior executives at Falcon Asset Management, the Center for Adaptive Systems Applications, and State Street Bank; professors with expertise in finance, operations research, applied mathematics, and financial engineering from Columbia University, Cornell University, Princeton University, and University of Pennsylvania; and a representative of the U.S. Department of the Treasury.

IMPACT OF ACADEMIC RESEARCH ON INDUSTRIAL PERFORMANCE

We invite your responses to these questions, either in the form of general comments or as responses to the more specific questions below. Your responses will be used by our Panel as background information for our report but will not be quoted verbatim without seeking your explicit permission.

1. Briefly describe fundamental or significant academic research contributions to the financial services industry as defined by the panel? (If possible, please supply references to published information that outlines the contributions.) *Use additional sheets, as necessary.*

2. Overall, would you describe the impact of academic research on industrial performance in the financial services industry over the past 20 years as (Please put an X in one box):

☐ 1. very large
☐ 2. large
☐ 3. medium
☐ 4. small
☐ 5. very small/nonexistent

If your response is "very large," could you please identify the specific areas of research you consider has had such a big impact.

3. What is the role of academic *research* in educating people who work in your industry? (Please focus on university *research* activities, rather than university education generally.) *Use additional sheets, as necessary.*

4. What structural forms of university-industry collaboration lead to good results in your industry? An example of such a structure might be a discipline- or industry-oriented "center" that solicits industry sponsors for a collection of projects that span a varied research program. What seem to be the essential determinants of success of such structures? *Use additional sheets, as necessary.*

5. What are significant emerging trends or problems that the financial services industry will face in the future that could benefit from academic research? *Use additional sheets, as necessary.*

6. What changes are required, if any, in academic research if it is to be responsive to these industrial trends and problems? *Use additional sheets, as necessary.*

7. What single step could be taken by universities to enhance the impact of academic research on the industry? *Use additional sheets, as necessary.*

8. What single step could be taken by companies to enhance the impact of academic research on industry? *Use additional sheets, as necessary.*

9. What single step could be taken by government to enhance the impact of academic research on industry? *Use additional sheets, as necessary.*

10. Do you see any downside to enhanced university-industry research collaboration? Things to be avoided? *Use additional sheets, as necessary.*

11. Other comments? Any comments, pointers to other studies, or suggestions would be appreciated. *Use additional sheets, as necessary.*

Worshop Agenda

ENHANCING ACADEMIC RESEARCH CONTRIBUTIONS TO THE FINANCIAL SERVICES INDUSTRY

October 15, 1998

National Academies Building
2101 Constitution Avenue, N.W.
Washington, D.C.

9:00 a.m. **Chairman's Opening Remarks and Self-Introductions**
Colin Crook, Former Chief Information Officer, Citicorp

9:15 a.m. **Introduction to the Task of the Financial Services Panel and Description of the Wider NAE Effort**
Colin Crook

10:00 a.m. Break

10:15 a.m. **Panel Discussion of the Financial Services Innovation System: Understanding the Process, Players, and Trends**
Chester Spatt, Mellon Bank Professor of Finance, and Director, Center for Financial Markets, Carnegie Mellon University
William W. Lang, Deputy Director for Special Studies, Office of the Comptroller of the Currency
Joe Dauber, Vice President, Customer Management, Novus Services, Inc.

12:00 p.m. Lunch in Meeting Room

12:30 p.m. **Panel Discussion of the Contributions of Academic Research to Financial Services: Past, Present, and Future**
Martin Holmer, The Policy Simulation Group
Paul Glasserman, Professor, Graduate School of Business, Columbia University
Sholom Rosen, Vice President, Citicorp

2:00 p.m. Break

2:15 p.m. **Panel Discussion on Assessing the Impact of Academic Research**

Jack Triplett, Brookings Institution

Patrick T. Harker, Department of Operations and Information Management, The Wharton School, University of Pennsylvania

Colin Carlton, Chief Investment Officer, Canada Trust Investment Management Group

3:45 p.m. **Academic Research and Financial Services: Where Are We Going?**

Colin Crook, Chairman

5:00 p.m. **Adjourn**

Workshop Attendees

Colin Crook, *Chair**
Chief Information Officer (retired)
Citicorp

John Alic
Lecturer in Energy, Environment, Science, and Technology
Washington, D.C.

Colin G. Carlton
Chief Investment Officer
Canada Trust Investment Management Group

Joe Dauber
Vice President, Customer Management
Novus Services, Inc.

John R. Davies
Chairman
Center for Adaptive Systems Applications

Judith M. Farvolden
Director, Research Communications
Algorithmics, Inc.

Paul Glasserman
Professor
Columbia University

Patrick T. Harker
UPS Professor and Chairman, Operations and Information Management Department
The Wharton School
University of Pennsylvania

Martin Holmer
President
Policy Simulation Group

William W. Lang
Deputy Director for Special Studies
Office of the Comptroller of the Currency

Blake D. LeBaron*
Center for Biological and Computational Learning
Massachusetts Institute of Technology

Deborah Malins
Senior Vice President
Product Development and Information Research
State Street Corporation

Alex Meeraus
President
GAMS Development

*Panel member

John M. Mulvey*
Department of Civil Engineering and Operations Research
Princeton University

Mitchell Rachlis
Senior Economist
General Accounting Office

Sholom Rosen
Vice President
Citibank

Chester S. Spatt
Mellon Bank Professor of Finance and Director, Center for Financial Markets
Carnegie Mellon University

Morris Tanenbaum*
Retired Vice Chairman and CFO
AT&T

Eric Thorlacius
Vice President
Falcon Asset Management

Jack E. Triplett
Brookings Institution

Andrew B. Whinston
Professor and Director of the CREC
College of Business Administration
University of Texas at Austin

National Research Council Staff

Stephen A. Merrill, Director, Board on Science, Technology, and Economic Policy

Thomas S. Arrison, Senior Staff Officer, Government-University-Industry Research Roundtable

NAE Program Office Staff

Tom Weimer, Director
Proctor Reid, Associate Director
Robert Morgan, NAE Fellow and Senior Analyst
Penelope Gibbs, Administrative Assistant

*Panel member

7

The Contributions and Impact of Academic Research

Significant contributions have been made by academic researchers in all five of the industries examined in this study. Each industry illustrates a unique pattern of industry-university research collaboration and different ways academic contributions are used. The study also reveals some crosscutting areas of academic research that are important to overall industrial performance. In addition, the mechanisms by which academia contributes have changed as traditional patterns of industry-university interaction—such as contract research, cooperative research, and personnel exchanges—have been augmented by new modes of interaction. Industry provides more than financial support for academic research, and academic research contributes more than technological advances to industry, although some contributions are difficult to measure in dollars. A comparison of patterns of interaction reveals a number of ways industry, academia, and government could realize even greater benefits through university-industry interaction.

PATTERNS OF UNIVERSITY-INDUSTRY INTERACTION

Finding 7-1. The nature of university-industry interactions varies from industry to industry as well as among companies within a given industry and individual academic institutions.

Each of the industries studied has a distinctive environment and poses different challenges for university researchers. In building infrastructure for network systems, universities have historically been test beds for new concepts and capabilities. For the medical devices and equipment industry, fundamental multidisciplinary research involving physical sciences and engineering, combined with academic medical centers, provides a critical environment for researching, developing, testing, and improving devices and for conducting the clinical trials necessary to obtaining regulatory approval, all in an atmosphere of close industry-university collaboration. In the aerospace industry, the mature, highly concentrated airframe, propulsion, and launch-vehicle sectors have a fairly narrow range of interactions with academic research, often using consulting agreements and contract research to develop better process methodologies and tools. By contrast, the less mature unmanned aerial vehicles sector of the industry looks to academic research for technical support, as well as for new concepts and understanding. In the transportation, distribution, and logistics services industry and the financial services industry, a sizeable cultural gap remains between industry and R&D in general, and academic research in particular. As a result, although academic research has had a significant impact on both, neither industry has developed interfaces with academic research comparable to those of the medical devices and equipment industry or the network systems and communications industry.

The wide variety of university-industry research interactions in these five industries makes it difficult to make generalizations. With the notable exception of multicompany research centers at universities, most financial support by industry is negotiated company by company. Companies have different needs and abilities to interact with academic researchers, and universities have different resources to devote to research of value to industry. Generalizations about what works best for all industries and universities should, therefore, be made very cautiously.

THE NATIONAL INNOVATION SYSTEM

Finding 7-2. The academic research enterprise is a major component of the national innovation system in the United States. The core competencies of academic research help sustain and leverage innovation to the benefit of industry.

Box 7-1 summarizes the innovation systems[1] for the five industries, which are innovative to varying degrees and in different ways that tend to change over time. R&D is only one element of their innovation systems but can be very important. Overall, the U.S. innovation system has many strengths:

- Open labor markets allow technically trained and educated people to move relatively freely between and among universities and industry.
- Reasonably robust review processes, coupled with a variety of support mechanisms ranging from peer-reviewed government grants to venture capital decisions, help to maintain high-quality research.

BOX 7-1
Industry Innovation Systems

Network Systems and Communications. This sector has a well developed innovation system that fosters the rapid creation and implementation of innovations, a strong research culture, and an industry structure that promotes innovation. Major growth segments are not impeded by excessive regulation, but allocation of radio-frequency spectrum may become increasingly contentious. The industry is heavily market and technology driven. Government and industry invest substantial funds in R&D relevant to the industry.

Medical Devices and Equipment. This industry has a well developed innovation system that benefits from broad public interest and support, including support from nonprofit organizations. It is probably the most tightly regulated of the five industries and the most affected by government policies related to safety, efficacy, and payments. Government and industry invest substantial funds in R&D relevant to the industry.

Aerospace. Most innovation in the larger, mature sectors is directed toward incremental improvements and cost reductions. Government support for R&D has been reduced in recent years, and the industry has undergone significant consolidation. The smaller, less mature sectors (unmanned aerial vehicles and space-based information services) are characterized by rapid innovation.

Transportation, Distribution, and Logistics Services. The innovation system is not as well established as in the network services and communications and the medical devices and equipment industries. Although historically this industry has not focused on R&D or technology-driven innovation, developments in integrated logistics are creating new opportunities for R&D. However, government and industry support for R&D in integrated logistics is modest.

Financial Services. An organized innovation system is slowly emerging as the result of rapid improvements and changes in information systems and technology. Overall, R&D is rudimentary and not broadly supported. Research opportunities in financial services underscore the need for a broad research portfolio that addresses both business and technological aspects of the industry drawing on social and behavioral sciences as well as engineering and applied mathematics.

- A large number of research structures and mechanisms, both internal and external to the university (e.g., academic departments, research centers, industry laboratories, start-up companies), provide multiple pathways for the commercialization of new ideas.
- A strong market and consumer demand for new technologies provides strong commercial incentives for introducing new technologies and, therefore, strong incentives for funding research to create them.

The research culture in the United States fosters innovation by supporting the movement of ideas and people among a broad range of diverse research sectors and structures. Even though research is often essential to innovation, there is rarely a linear progression from a research result to advanced development to product development to economic return. Ideas and people tend to bounce around, and new ideas are sometimes stymied by political or business impediments and forced to find alternative routes to implementation.

CONTRIBUTIONS OF ACADEMIC RESEARCH

Contributions of academic research to the five industries studied include: graduates trained in modern research techniques; fundamental concepts and "key ideas" resulting from basic and applied research; and the development and testing of tools, prototypes, and marketable products, processes, and services. The sources of these contributions include engineering, the natural sciences, computer sciences, mathematics, social sciences, behavioral sciences, management studies, and policy sciences.

Graduates Trained in Research

Finding 7-3. University-based research provides an education/training ground for entrants into the industrial workforce. Integrated research and education helps maintain the flow of human resources from universities that contributes to an educated, trained industrial workforce. University graduates and faculty are also involved in many technology-based, start-up companies.

Students trained in research are a major component of academia's contribution to industrial performance. U.S. universities are the primary source of people with research training and experience, including undergraduates, graduate students, postdoctoral researchers, and faculty. Individuals with research training are highly valued by industry, whether or not they are involved in research for the companies that hire them. In addition to the specific body of knowledge acquired through academic research, industry values research experience because it requires abilities that are prized in any technical endeavor: self-motivation, problem solving, teamwork, an understanding of related research, contacts with other researchers and colleagues, the ability to organize material, and the ability to overcome setbacks. Research-trained industry employees also enhance a company's capacity to absorb new ideas, including the results of research, even if the company does not conduct its own research.

Academic researchers also participate in new companies. Many technology-based start-up companies emerge from academic research and continue to attract research graduates as they grow. Many high-technology clusters around the country have developed around one or more research university.

Contributions from Basic, Long-Term Research

Finding 7-4. Contributions from basic, long-term academic research in a broad spectrum of disciplines have figured prominently in industry performance.

Portfolio theory, linear programming, derivative-pricing theory, and prospect theory, all of academic origin, have laid the foundation for whole new families of financial products and services. Academic contributions to linear and integer programming and queue theory are the building blocks of the information-management and decision-support technologies at the heart of the integrated-logistics revolution. Medical devices, such as magnetic resonance imaging machines and pacemakers, are based on the contributions of fundamental research from multiple disciplines in the natural sciences and engineering. In the network systems and communications industry, universities have made important research contributions to the development of digital subscriber-line technology, third-generation wireless communication, computer graphics, databases, search engines, generalized processor sharing, parallel processing, traffic management, and stable broadcast networking. In aerospace, contributions of basic research include: the theoretical basis for flight controls for unmanned aerial vehicles; Shannon's information theory (e.g., as applies to communication with aircraft, spacecraft, and satellites); electromagnetic antenna theory; linearized unsteady-flow analysis; composite-laminate theory; improved understanding of fiber-matrix interactions in composite materials; superplasticity; and real-time decision systems using artificial intelligence.

Basic, long-term research is essential to the university's role as creator of new knowledge and understanding. The committee's review of these five industries confirms that the results of basic research in a wide range of disciplines eventually find their way by diverse paths into many aspects of commercial life.

Contributions from Applied Research

Finding 7-5. Academic researchers in applied research and the academic research infrastructure are directly involved in the development of industrial tools, prototypes, products, and production processes, as well as the delivery of products and services.

The five industries in this study provide a variety of examples of the contributions of applied research (see Box 7-2). Sometimes applied research is protracted and has cumulative, incremental results. An example might be continued improvements in computational fluid dynamics as a tool for modeling airflow. Another would be the long-term contributions of academic researchers to improved production processes and product performance in electronic storage devices.

BOX 7-2
Contributions of Applied University Research

Network Systems and Communications. Contributions include packet switching and the Internet TCP/IP protocol, both key elements in the development of the Internet. The Mosaic web browser interface was an important step in the rapid evolution of the World Wide Web. University researchers and other university personnel have contributed in significant ways to routers, the development of ATM switches, digital subscriber-line technology (DSL), third-generation wireless transmission, computer graphics, search engines, traffic management, stable broadcast networking, the evolution of new networks, and the development of standards.

Medical Devices and Equipment. The development of a wide range of therapeutic and diagnostic devices has resulted from the involvement of academic researchers and academic medical centers in R&D, prototype testing, evaluation, and clinical trials. Devices and equipment include magnetic resonance imaging equipment; whole-body CAT scanners; flexible endoscopy; lasers for a broad range of medical applications, ranging from gastrointestinal surgery to eye surgery; cardiac-assist devices; organ and joint replacements; ultrasound and minimally invasive surgical techniques; and developments in tissue engineering.

Aerospace. Contributions to the development of tools include advanced nonintrusive instrumentation, flow-visualization techniques, and computational fluid dynamics. Contributions to specific technologies have been made in the areas of heat transfer, combustion cooling, and aeromechanics; low-Reynolds-number airfoil design; Internet by satellite, including protocols and computational tools for data integration; and folding-wing design for small unmanned aerial vehicles.

Transportation, Distribution, and Logistics Services. Contributions include optimization modeling for shippers, software applications/decision-support systems for routing, production scheduling, logistics, and distribution management. Much of this software was commercialized by academic spin-off companies.

Financial Services. Contributions include new financial instruments (including index funds and derivatives); financial information and research tools, including risk/credit metrics and financial risk-management software; models for pricing derivatives and securities; and advances in cryptography for specific financial services. These contributions have come from engineering and business schools, as well as from mathematics research.

The ability to solve discrete practical problems is also valuable in countless projects performed for individual companies. Short-term research projects, student projects, and consulting projects to solve specific, important problems in industry are based on formal and informal relationships between companies and faculty. Many companies nurture relationships with multiple universities, often relying on local institutions to solve technical problems or to advise the company's engineering staff on potential solutions. Examples include assistance in production scheduling in logistics, simulations of airflow and nondestructive evaluation of materials in aerospace, and models for pricing derivatives and securities in financial services. Most universities consider this an important aspect of the service role of the university and encourage these interactions.

Research centers, especially those with industrial participation, are another avenue by which universities perform both "directed" basic research and applied research that helps industry. In 1993, almost 70 percent of industry's financial support for university research flowed through some 1,100 university-industry research centers, which have become the dominant form of industry support for academic R&D (Cohen et al., 1998). The best known examples are the engineering research centers (ERCs) funded by the National Science Foundation (NSF) since the mid-1980s. NSF currently funds 20 ERCs in four broad categories: bioengineering; design, manufacturing, and product development systems; earthquake engineering; and microelectronic systems and information technology (NSF, 2001). Industry participation and industry's help in defining problems that are of interest to many companies, have greatly increased the impact of academic research centers.[2]

Key Ideas

Finding 7-6. Sustained interactions between academic research and industry have been a source of "key ideas" that have generated significant technological opportunities through a fusion of knowledge of the possible and knowledge of what needs to be done.

Specific contributions of academic research—basic and applied—represent key ideas derived from sustained interaction between academic research and industry. Key ideas and the major technological opportunities or breakthroughs that result from key ideas are often the product of cumulative research interactions and advances involving the flow of ideas and people back and forth across the boundaries between universities and industry. Examples of key ideas include: the TCP/IP Internet protocol, the web browser, routers, index funds and derivatives, decision-support technologies, pacemakers, and magnetic-resonance imaging. (For a graphic illustration of the interaction between academic and industry research on key ideas leading to major technological advances in information technology see Figure 2-1.)

Contributions from Multi-industry, Indirect, and Complementary Research

Finding 7-7. Academic research in a given field or discipline may contribute directly or indirectly to more than one industry; and many innovations result from complementary advances in more than one field of research.

Many contributions of academic research to an industry are mediated through other disciplines or embedded in technologies, products, and services derived from other industries. Basic research in physics, biology, and chemistry has led to new knowledge and capabilities in microelectronics, genetic engineering, and other fields that have directly contributed to the creation of high-value, high-technology products and services. Contributions from academic research to major cross-sector technologies, such as information technology, have directly benefited many industries. For example, information technology is critical to the technical and market performance of aircraft and has profoundly changed the structure and performance of the financial services, as well as the transportation, distribution, and logistics services industries. Similarly, intelligent sensors, computer-aided diagnosis, and robotics are the basis for many new medical devices. Research in materials science and bioengineering has enabled advances in products and processes in many industries.

The five industry studies have also underscored the multidisciplinary character of many innovations in products and services. For example, the development of new medical devices relies heavily on advances not only in the life sciences, but also in the physical sciences and engineering. Many service innovations in the network systems and communications industry have depended on complementary progress in engineering and physical, social, and behavioral sciences.

Contributions from the Social Sciences

Finding 7-8. Many valuable contributions to industry have resulted from academic research in the social, behavioral, management, and policy sciences.

Network Systems and Communications Industry

Academic business schools have long been concerned with making the benefits of information technology available to businesses. Through research, a number of approaches and techniques have been developed, including decision-support systems, the implementation of information technology for strategic advantage, computer-supported cooperation, productivity research, and software development methodologies. Deregulation, partly a response to academic research in economics, has affected all five of the industries in this study to varying degrees. Economics research on network externalities and Internet economics has helped to define business strategies for electronic businesses and Internet service providers. Organizational aspects of communications service companies are the focus of attention in new information-management schools and programs. Research by psychologists and social scientists has explored how people use computer and communication systems and the effects of these systems on people and organizations. An excellent example is a classic study by Card et al. (1983) showing how cognitive psychology can be used to estimate human performance when interacting with a computer.

Medical Devices and Equipment Industry

Research in engineering, the natural sciences, the social sciences, risk analysis, and business are likely to be of increasing importance to medical information systems. In addition, clinical research studies that help determine the acceptance or rejection of new medical devices require a broad-based approach that incorporates a variety of disciplines.

Aerospace Industry

In a broad sense, academic research on production and management systems, typically performed by business faculty, has had an enormous impact on all manufacturing industries. Concepts that have contributed to recent increases in productivity in manufacturing, such as total quality management, workforce empowerment, supply-chain integration, and just-in-time production, were identified and disseminated by academic researchers. The Lean Aerospace Initiative (LAI) at the Massachusetts Institute of Technology is an example of multidisciplinary research with strong industry participation. Focused on strategies for applying lean manufacturing and management concepts to aerospace, LAI includes research in several engineering disciplines, economics, behavioral science, computer science, marketing, management, and other disciplines.

Transportation, Distribution, and Logistics Services Industry

The research most relevant to integrated logistics is in operations research, an area associated with engineering schools, applied mathematics departments, and business or management programs. Human factors research and consumer research are also important to the industry.

Financial Services Industry

The financial services industry has a history of benefiting from economics and business research rather than research in natural sciences and engineering. Leading examples are the Nobel prize-winning work in economics by Markowitz and Sharpe on portfolio theory, by Scholes and Merton on pricing derivative securities, and by Koopmans and Kantrovich on linear pricing models. Academic research in the social sciences has contributed to an understanding of large data sets and consumer behavior. In addition, academic researchers have been instrumental in analyzing legal and regulatory restraints on financial services.

VECTORS OF CONTRIBUTION

Finding 7-9. Numerous diverse, robust, and often mutually reinforcing vectors link academic research to the five industries.

The traditional idea of universities as places that educate students and conduct basic research is, at best, incomplete. Ideas and people are carried by multiple vectors between academia and industry. These vectors include the direct hiring of students, graduates, and faculty by industry; temporary exchanges of researchers; faculty consulting arrangements; sabbaticals; research grants and contracts; institutional mechanisms at universities (e.g., research centers, consortia, industrial liaison programs); technology licensing; spin-off companies; publications; conferences; and short courses. The modes or pathways are summarized in Box 7-3.

IMPACT OF CONTRIBUTIONS

Measuring the quantitative impact of specific innovations on the performance of a firm or an industry is extremely difficult because performance in the market is determined by synergies between multiple innovations and other factors, both internal and external. Isolating the contribution of academic research is at best an inexact science. Therefore, this study was designed from the outset to provide a qualitative assessment. Panels for each industry relied on informed opinion, informal surveys of industry and academic leaders, workshop discussions, and expert judgment to assess the impact of academic research on industry performance.

Network Systems and Communications Industry

Academic research has had a substantial impact on this industry. The flow of researchers, ideas, and entrepreneurial activity between universities and industry, coupled with government support for research and test beds for infrastructure development, have been instrumental in the creation of new companies, services, and modes of business. As emphasis shifts to the deployment and maintenance of large-scale systems and the economical provision of services, the impact of university research may be moderated somewhat because research relevant to operational networks is expensive and often proprietary. Many firms opt to hire capable university researchers, rather than fund research at universities. However, the federal government continues to invest heavily in academic research on information technology, which is expected to generate results with long-term commercial impacts.

Medical Devices and Equipment Industry

Academic research has had a substantial impact on performance in this industry. In addition to science- and technology-based research and innovation in universities, academic medical centers (AMCs) play a unique role. Industry and academia depend on each other for product development, testing, introduction,

BOX 7-3
Vectors of Technology Transfer from Universities to Industry

Trained Graduates. The direct employment of graduates and faculty, whether as permanent or temporary hires, by manufacturing companies and by the intermediate product and service vendors that support them is an important vector of knowledge/technology transfer from academic research to industry.

Consultancies. Faculty consultancies are not well documented but represent an important means of exchanging knowledge. Data on the amount of time faculty spend consulting with industry are not available. Most engineering schools, however, allow faculty to spend up to 20 percent of their time as consultants.

Sponsored and Collaborative Research. Research grants, primarily from the federal government but also from other nonindustry sources, can lead to results that find their way to industry. Research funded directly by industry accounts for roughly 8 percent of total academic research activity; industry support is sometimes a requirement for federal funding.

Spin-off Companies. Companies started by academic researchers have become an increasingly important mechanism of knowledge/technology transfer in the last two decades.[1] CISCO Systems, Amati Communications, and Growth Networks in the network systems industry and CPLEX Optimization and CAPS Logistics in the transportation, distribution, and logistics industry are examples of the successful commercialization of the results of academic research by start-up companies.

Licensing. Technology licensing directly from the university or through spin-off companies increased rapidly during the 1990s. University gross revenues from licensing increased from $130 million in 1990 to a total of $1.1 billion in 2001 (AUTM, 2002; NSB, 2000).[2] Patent ownership by universities rose sharply, from about 250 annually in the early 1970s to more than 3,700 in 2001 (AUTM, 2002; NSB, 2000). However, royalties from university patents represent only about 2 percent of R&D expenditures and generally come from a few blockbuster patents (Nelson, 2000). Patenting and licensing by universities have been mostly in the biotechnology sector (NSB, 2000).

Research Centers. The number and importance of university-industry research centers has increased substantially in the last two decades. In general, these relationships appear to play a useful facilitating role in university-industry interaction and knowledge transfer.

Publications, Conferences, Workshops, and Short Courses. Research publications and the presentation of research results at conferences, workshops, and other forums that bring together academic and industrial researchers remain important vectors for the exchange of knowledge between academia and industry.

[1] More than 3,870 new companies were formed based on licenses from academic institutions between 1980 and 2001, 494 of them in 2001. In that year, universities held an equity interest in 70 percent of these start-up companies. Many more companies based on university-generated ideas, science, and technology did not involve university licensing (AUTM, 2002).

[2] These figures were self-reported by respondents to the AUTM Licensing Survey and do not necessarily include all universities (AUTM, 2002).

and modification of medical devices at AMCs. Nevertheless, the impact could be even greater if there were a more systematic approach to educational partnerships between industry and universities. These partnerships might include the sharing of large, expensive medical research facilities and joint research and training activities.

Aerospace Industry

Although many basic concepts and a good deal of the fundamental knowledge of aerodynamics were developed by academics, the impact of academic research on performance in the mature sectors of the aerospace industry has been relatively modest in recent years. The combination of aerospace corporations' strong research capabilities, concerns about intellectual property rights, an emphasis on incremental improvements in products with long life cycles, and competitive pressures that demand improvements in production processes has limited the current value of academic research. However, industry and government laboratories that support the mature aspects of the industry continue to depend on graduates educated and trained at universities, and aerospace continues to be one of the most research-intensive industries. Less mature sectors, such as unmanned aerial vehicles and space-based information systems, where innovation is proceeding rapidly, depend heavily on universities for research and innovation.

Transportation, Distribution, and Logistics Services Industry

The impact of academic research on performance in transportation, distribution, and logistics services as a whole has been relatively modest. Basic research, some of it done in the 1950s with no logistics applications in mind, has had the greatest impact on the industry. Linear and integer programming and queueing theory are the building blocks of the information-management and decision-support technologies at the heart of the integrated-logistics revolution. Applied research in these fields has also made important contributions, such as large-scale optimization modeling, decomposition methods, network optimization, and advances in other areas of operations research. These successes notwithstanding, a large gap remains between technologies that could have a tremendous impact and technologies that actually have had an impact. Overall, there is not enough demand from the industry or capacity in the industry to take advantage of research results.

Financial Services Industry

Academic research has had a substantial impact on certain aspects of the financial services industry, especially on novel financial products and services. The financial instruments, investment management, and decision-support tools created in academia or through public-private partnerships have been critical to new products, business lines, and more efficient transactions, and hence to the success of the industry. Furthermore, a significant component of modern financial services would not be possible without the fundamental mathematical tools developed for or adapted to financial problems by academia. Research has focused on linear programming models in economics, portfolio theory, and pricing derivative securities. University research has greatly impacted financial services despite the lack of a well developed, organized R&D system focused directly on problems and issues related to financial services.

NOTES

[1]Drawing on Lundvall's (1992) definition of national innovation systems, the committee defines an industry's system of innovation as "constituted by elements (people, capital, organizations, et al.) and relationships that interact in the production, diffusion and use of new, and economically useful, knowledge" within a given industry.

[2]At the end of the funding period for NSF-supported engineering research centers, most ERCs will become self-sustaining. Currently there are 16 self-sustaining ERCs.

REFERENCES

AUTM (Association of University Technology Managers). 2002. AUTM Licensing Survey, FY 2001. Northbrook, Ill.: AUTM.

Card, S.K., T.P. Moran, and A. Newell. 1983. The Psychology of Human-Computer Interaction. Hillsdale, N.J.: Lawrence Erlbaum Associates.

Cohen, W.M., R. Florida, L.P. Randazzese, and J. Walsh. 1998. Industry and the Academy: Uneasy Partners in the Cause of Technological Advance. Pp. 171–199 in Challenges to Research Universities, R. Noll, ed. Washington, D.C.: The Brookings Institution.

Lundvall, B.L. 1992. National Systems of Innovation: Towards a Theory of Innovation and Interactive Learning. London: Pinter Publishers.

Nelson, L. 2000. Many Forms of Technology Transfer from Universities. Viewgraph presented at the AAAS Science and Technology Policy Colloquium, Washington, D.C., April 2000.

NSB (National Science Board). 2000. Science and Engineering Indicators 2000. Washington, D.C.: U.S. Government Printing Office.

NSF (National Science Foundation). 2001. Engineering Research Centers: Partnerships for Competitiveness. Arlington, Va.: National Science Foundation.

8

Challenges for the Future

Based on the conditions, experiences, and trends in the five industries examined in this study, the committee identified six major challenges that are likely to affect the impact of university-based research on industry performance in the coming years.

SERVICES

Finding 8-1. Although innovations in service delivery are becoming more important, the academic research enterprise is not focused on or organized to meet the needs of service businesses.

Services account for almost 80 percent of the U.S. gross domestic product, employ a large and growing share of the science and engineering workforce, and are the primary users of information technology. In most manufacturing industries, the service functions, such as logistics, distribution, and customer service, have become leading sources of competitive advantage. The rate of innovation and level of productivity in the services infrastructure (e.g., finance, transportation, communication, health care) have an enormous impact on the productivity and performance of all other segments of the economy. Moreover, as the studies of the financial services industry, the transportation, distribution, and logistics services industry, and the network systems and communications industry show, improving services is a major impetus for innovation throughout the economy.

Nevertheless, the U.S. academic research enterprise, despite its broad disciplinary base and potential for crossdisciplinary research and training, is not focused on or organized to meet the needs of service businesses. Major challenges facing universities are: (1) adapting and applying systems and industrial-engineering concepts, methodologies, and quality-control processes to service functions and businesses; (2) integrating technological research with research in social sciences, management, and public policy; and (3) educating and training engineering and science graduates to deal with management, policy, and social issues.

THE REGULATORY CLIMATE

Finding 8-2. Regulations and regulatory changes have profoundly influenced industry receptivity to contributions by academic research. In some cases, academic research has helped to shape the regulatory environment; academia is well qualified to provide interdisciplinary expertise to inform regulatory decisions.

All of the industries in this study operate in an environment that is currently or has been highly regulated, and changes in the regulatory environment over time have affected them in obvious and not so obvious ways. At one end of the spectrum are the airline and trucking industries. For many years, strict regulation precluded airlines and trucking companies from competing based on price; therefore, competition was based on speed, reliability, and other amenities that tended to spur innovation. Deregulation in both industries has led to intense competition, based on both price and quality of service. Although the need for innovation remains, lower margins and, therefore, tighter research budgets have restricted the focus of R&D. At the other end of the spectrum is the medical devices and equipment industry. Companies in this industry operate in a highly regulated environment in which the safety and effectiveness of new devices must be clearly demonstrated. Universities, specifically academic medical centers, are particularly well equipped to carry out the laboratory research and clinical trials described in regulatory requirements. This industry is a prime example of how the contributions of academic research are affected by the overall regulatory environment in which an industry operates.

Academic research has also greatly influenced the regulatory environment. Based on economics research, much of it performed in academia, the role of regulation has been redefined from protecting the public interest in naturally monopolistic markets to promoting market entry and ensuring vigorous competition to achieve public benefits. The change has spurred deregulation in a number of industries, including network systems and communications. In the financial services industry, the impact of academic research on regulation has been small historically, usually in response to crises; however, the impact is growing, especially in the area of risk management. The influx of technically trained scientists and engineers into financial regulatory bodies has enabled regulators to draw on advances in risk modeling, which, in turn, has led to innovations in the industry proper. In the medical devices arena, academic researchers could play an important role in the ongoing reform of the Food and Drug Administration's regulatory policies by assembling industry, regulatory, and clinical panels to discuss appropriate requirements for bringing products, such as artificial hearts and mechanical cardiac-assist devices, into widespread clinical use.

INTELLECTUAL PROPERTY

Finding 8-3. Over the past 25 years, research universities have increasingly emphasized technology transfer and the generation of income from research results, through patenting, the creation of technology transfer offices, and licensing. Although the increased attention to management of intellectual property has had many positive consequences for industry and academia, questions remain about the overall effectiveness of technology transfer investments, as well as the impact of these activities on universities' core research and education missions.

Throughout the 1970s and early 1980s, government policy increasingly favored stronger protections for intellectual property resulting from publicly funded research. Several universities had already increased patenting activity in the 1970s, largely as a result of the emergence of biotechnology, but the propensity to patent increased markedly after 1980. Since 1980, the number of university technology transfer offices has grown from 25 to more than 200 (Sampat and Nelson, 1999). These offices have provided an alternative interface with industry to the traditional offices of sponsored research. Technology-transfer offices focus on licensing university technologies and generating royalties.

Other mechanisms for profiting from research have also been developed. For instance, some leading research universities have made long-term agreements with individual companies for joint research, joint clinical trials, and profit sharing; companies use these agreements to leverage research funded from other sources. Universities are becoming more willing to take equity stakes in new companies. A percentage of equity is often a requirement for companies' participation in university-based business incubators, and universities are increasingly providing some of the initial funding for faculty members' start-up companies. In addition, a growing number of university researchers have added financial gain and entrepreneurship to their traditional university roles of teaching, research, and service. In some cases, patents and royalties are shared with the university. In other cases, faculty researchers have taken advantage of available venture capital to fund new companies to produce commercial products based on their research. All of these mechanisms—patents, licensing, contracting, industrial liaison programs, and start-up companies—have expanded interactions between universities and industry and changed the traditional role of universities in that relationship, which, for the most part, had been limited to faculty consulting, small amounts of contract research for companies, and the preparation of graduates for careers in industry.

The emphasis on commercializing research results has affected the structure and dynamics of industry interactions with leading research universities. For example, in the mature sectors of the aerospace industry the problem of the protection and ownership of intellectual property is a significant barrier to collaborative university-industry research. This situation may be attributable to the advanced level of competition in the industry and could be indicative of things to come in other mature, highly competitive, research-intensive industries. By contrast, the treatment of intellectual property rights has not figured prominently in relations between universities and the financial services industry, in which secrecy and first-mover advantages prevail. Indeed, the lack of intellectual property rights in financial services might be an impediment to collaborative academic-industry research in a different way because it is difficult for universities to maintain secrecy. In the transportation, distribution, and logistics services industry, where intellectual property is often developed and commercialized by the same faculty, consultancies and start-up software companies have helped the industry avoid problems.

Increased patenting activity by academic researchers has had many positive consequences for both industry and academia. Academic researchers now have new incentives and new avenues for pursuing their entrepreneurial energies and new products, services, processes, and companies to show for it. As Congress predicted when it passed the Bayh-Dole Act in 1980, allowing universities to own and profit from the results of their research has stimulated researchers to patent and seek the commercialization of research results. Recent research indicates that the willingness of industry to invest in the commercialization of inventions licensed from universities is closely correlated with strong property rights (Jensen and Thursby, 2001; Dechenaux et al., 2002). The increase in patents has also had benefits for industry: (1) patenting places research results in the public domain, where they are much more accessible than they are through journals or unpublished papers; (2) patents have value that can be capitalized through licensing agreements or as collateral in securing financial resources for start-up companies; and (3) patents provide at least some protection for the resulting commercial products, thereby encouraging investors to make the capital investments necessary for successful commercialization.

At the same time, the growing emphasis on university ownership and exploitation of intellectual property has raised questions about the near-term efficacy of the patent-licensing infrastructure, as well as the long-term impact on university-industry interactions and the health of the academic research enterprise. To date, direct contributions of academic research through patenting activity has been small. The creation of technology-transfer offices, with the technical, financial, and legal expertise they require, however, can be expensive; indeed, relatively few universities have earned much of a direct financial return on these investments. In fiscal year 2000, only 72 universities had income from licenses of more than $1 million; and the University of California system accounted for nearly one-quarter of the total license income reported in that year. Although even small amounts of income are important in the context of tight university budgets, license income exceeded 5 percent of total research expenditures at only 15 universities; typically it is less than 1 percent (AUTM, 2001).

Students of technology transfer and the research university have begun to look into questions raised by the growing interest of universities and research faculty in intellectual property (e.g., Henderson et al., 1998; Link et al., 2002; Morgan and Strickland, 2000; Nelson, 2001; Press and Washburn, 2000; Stefan, 2000). Have investments in patent licensing infrastructure been worth it? Have they made technology transfer from universities to industry more effective? Or would the licensed technologies have been picked up by industry in any case? Does the emphasis on capturing intellectual property rights raise the transaction cost of research? Are there other value-added results from these investments, and how should they be measured? To what extent does the emphasis on intellectual property and secrecy inhibit the free flow of ideas, impede advances, and disrupt the research culture? How has increased emphasis on intellectual property affected teaching and learning? A better understanding of these and related issues will have important implications for future practices and policies. Therefore, it is critical that these questions continue to be addressed.

INFORMATION TECHNOLOGY

Finding 8-4. Information technology is critical to the performance of all industries and will continue to be so in the future. Industry's need for the continued development, diffusion, and application of advanced information technology presents major opportunities for academic research in many disciplines, including mathematics, computer sciences, physical sciences, life sciences, multiple engineering disciplines, social sciences, and behavioral sciences.

The importance of information technology to industry performance cannot be overstated. As hardware becomes cheaper and more powerful, networks and communications are becoming more pervasive, and the volume of information created, stored, and exchanged is growing exponentially. As a result, questions about the management of information for private gain and/or public benefit are also increasing. Addressing these issues will provide a wide range of challenges for academic researchers in almost every discipline. To name just a few examples, industries will need software that facilitates the interoperability of legacy systems and reduces the vulnerability of infrastructure and business to breaches of security and privacy. Academia will also be expected to continue supplying skilled technicians, developers, and managers (NRC, 2000).

A BALANCED RESEARCH PORTFOLIO

Finding 8-5. Universities must maintain a balance of research projects to sustain their role as repositories of expertise and resources in many disciplines—basic and applied research in engineering, life sciences, physical sciences, social sciences, behavioral sciences, managerial sciences, public policy studies, and interdisciplinary research.

Basic, long-term research performed at universities is an essential part of the national innovation system. All of the industries studied derive significant benefits from basic research, although the importance of basic research to industry performance is not well recognized, particularly among individual companies. Most federal funding continues to support basic research, but industry-funded academic research is focused mostly on problems that can be solved relatively quickly. As universities become more entrepreneurial, the potential financial gains from commercially relevant research could create strong incentives for universities to focus on applied research at the expense of basic research. If so, federal funding for basic, long-term research will be even more critical.

Federal funding is now virtually the only source of support for basic research, which makes effective management of federal research programs of paramount importance. At the highest level, Congress should recognize and reaffirm the importance of basic research at universities. To capture the imaginations of the best academic researchers, program managers at the Defense Advanced Research Projects Agency, National Science Foundation (NSF), and other agencies should work with researchers to develop agendas that might lead to major new insights. In some areas, such as network systems, the best researchers may already be losing interest. The challenge is not just to maintain a balance between basic and applied research, but also to ensure that the basic research portfolio is sufficiently diverse to stimulate innovative thinking by academic researchers in many fields.

To meet this challenge, the balance of federal funding for research in specific fields and agencies should be reassessed. The percentage of federal funding for academic research supported by the National Institutes of Health increased from 49 percent to 62 percent from 1980 to 2001. During the same period, NSF's funding decreased somewhat, from 20 percent to 17 percent. The relative shares of federal funding for academic research at other federal agencies, such as the National Aeronautics and Space Administration, the U.S. Department of Energy, and the U.S. Department of Defense, have also declined (NSF, 2001). In a policy statement accompanying *Science and Engineering Indicators 2000*, the National Science Board noted (NSB, 2000):

> The life sciences now account for more than 50 percent of the U.S. federal investment in basic research.... Today's strong federal support for the life sciences is warranted because biomedical research is on the cusp of a revolution in preventative medicine and treatment. Nevertheless, today's overall research budget is increasingly out of balance.

The generation of key ideas that lead to technological breakthroughs, as well as sustained incremental innovation, requires contributions—both direct and indirect via cross-sector technologies—from research in many fields. The value of research results in one field often depends heavily on advances in complementary fields, which is a strong argument for maintaining a balanced research portfolio in many fields of science and engineering.

Finally, research opportunities in the social and behavioral sciences pose another challenge for industry program managers, as well as academic researchers. As the U.S. economy continues to shift toward services, and competitiveness in manufacturing and services industries is defined increasingly by the relative ability of firms to manage knowledge and human capital, as well as to anticipate and meet the wants of customers (involving them more and more directly in the design and production of goods and services), the importance of research in social, managerial, behavioral, and policy sciences for industry and government will certainly grow. For the most part, the value and relevance of this research has yet to be recognized by industry or government agencies. Although research in selected areas of economics and managerial sciences has had a demonstrable impact on industry practices, researchers in the social and behavioral sciences, in general, have not conveyed the value of their research to industry effectively. Examples in this study of five very different industries show that research in the social and behavioral sciences (integrated with the natural sciences and engineering) in areas related to information technology and services can greatly improve our understanding of how technological developments affect individuals and society as a whole.

KEEPING PACE AND MOVING FORWARD

Finding 8-6. The core strengths of the academic research enterprise are stable enough and flexible enough to respond to the rapidly changing needs of industry.

A major challenge for universities is keeping pace with the rapidly changing research and human resource needs of industries while continuing to pursue basic research in new areas to generate ideas that will provide the foundation for industries in the future. This challenge is manifested differently in different industries. In the network systems and communication industry, as well as the medical devices and equipment industry, where linkages to academic research have been very strong, industry leaders are concerned that academic research may not be able to adapt, articulate, and pursue basic and applied research and training in new directions. In the financial services and the transportation, distribution, and logistics services industries, which do not have a strong industry R&D ethos, and in the mature sectors of the aerospace industry, industry is less concerned about academic research "keeping up" than about academia meeting their research and educational/training needs. All five industry studies revealed problems with participants in academic long-term research adapting to shifting industry priorities and emerging opportunities in particular areas.

The academic research enterprise has some very important core strengths. Universities address a broader spectrum of ideas and disciplinary perspectives than any other institutions in the U.S. innovation system; they have enormous potential for multidisciplinary research. Universities also integrate advanced research and education. The constant flow of new students through universities continuously revitalizes the academic research enterprise, challenging the assumptions of faculty and bringing fresh perspectives to research. Research-trained graduates play a critical role in the development, transfer, diffusion, and application of new knowledge and technology in industry. Universities can draw on these core strengths to keep pace with current industry needs and move forward.

REFERENCES

AUTM (Association of University Technology Managers). 2001. AUTM Licensing Survey, FY 2000. Norwalk, Conn.: AUTM.

Dechenaux, E., B. Goldfarb, S. Shane, and M. Thursby. 2002. Appropriability and the timing of innovation: evidence from MIT inventions. Summer Institute 2002, National Bureau of Economic Research, Cambridge, Mass. Available online at: *http://www.nber.org/~confer/2002/si2002/thursby.pdf*. [June 24, 2003]

Henderson, R., A. Jaffe, and M. Trajtenberg. 1998. Universities as a source of commercial technology: a detailed analysis of university patenting, 1965–1988. Review of Economics and Statistics 80(1): 119–127.

Jensen, R. and Thursby, M. 2001. Proofs and prototypes for sale: the licensing of university inventions. American Economic Review 91(1): 240–259.

Link, A.N., J. Scott, and D. Siegel. 2002. The economics of intellectual property at universities: an overview (forthcoming in Special Issue of the International Journal of Industrial Organization). Available online at: *http://www.people.virginia.edu/~sa9w/ijio/abstracts/Accepted/Overview.pdf*. [June 24, 2003]

Morgan, R.P., and D.E. Strickland. 2000. U.S. university research contributions to industry: findings and conjectures. Science and Public Policy 28(2): 113–121.

Nelson, R.R. 2001. Observations on the post-Bayh-Dole rise of patenting at American universities. Journal of Technology Transfer 26(1-2): 13–19.

NRC (National Research Council). 2000. Making IT Better: Expanding the Scope of Information Technology Research to Meet Society's Needs. Computer Science and Telecommunications Board. Washington, D.C.: National Academies Press.

NSB (National Science Board). 2000. Science and Technology Policy: Past and Prologue. A Companion to Science and Engineering Indicators, 2000. Available online at: *http://www.nsf.gov/pubs/2000/nsb0087/nsb0087.pdf*. [June 24, 2003]

NSF (National Science Foundation). 2001. Federal Funds for Research and Development: Federal Obligations for Research to Universities and Colleges by Agency and Detailed Field of Science and Engineering: Fiscal years 1973–2001. Arlington, Va.: National Science Foundation.

Press, E., and J. Washburn. 2000. The kept university. Atlantic Monthly 285(3): 39–54.

Sampat, B., and R. Nelson. 1999. The Emergence and Standardization of University Technology Transfer Offices: A Case Study of Institutional Change. Prepared for the 1999 Conference of the International Society for the New Institutional Economics, September 16–18, 1999. Washington, D.C.: World Bank.

Stefan, P. 2000. Remarks at the Conference on University-Industry Technology Transfer, Purdue University, West Lafayette, Indiana, June 10, 2000.

9

Conclusions and Recommendations

This study has documented how performance in five very different industries has been improved by contributions from academic research—from the education of graduates with research training at all levels to conceptual breakthroughs and incremental technical advances based on basic and applied research to the development and testing of tools, prototypes, and marketable products, processes, and services. Numerous diverse, robust, and often mutually reinforcing vectors link academic research to industry, including direct hiring of students, graduates, and faculty; temporary exchanges of researchers; faculty consultancies; industry-sponsored research contracts and grants; research centers; consortia; industrial liaison programs; technology licensing; start-up companies; publications; and conferences.

Research-trained graduates at all degree levels are critical to the development, transfer, diffusion, and application of new knowledge and technology in industry. Indeed, the interaction between research-trained individuals in academia and in industry is essential to the exchange of knowledge and technology and to the nation's innovation system.

Basic long-term research at universities in many fields of science and engineering has had a huge impact on industry. Roughly half of all basic research in the United States is conducted by universities. Basic long-term research is essential to the health of universities as creators of new knowledge and understanding; and the results of basic research in a wide range of disciplines intermingle, build on each other, and eventually find their way by diverse paths into commercial life. Because in most fields it is difficult for individual firms or groups of firms to appropriate return on investment exclusively, commercial firms have little incentive to fund basic research at universities. As a result, the federal government currently funds the majority of basic research in academia.

Contributions from applied research at universities are also very important to industry. The industry studies in this report document that academic researchers and the academic research infrastructure are directly involved in the development of industrial tools, prototypes, products, and production processes, as well as in the delivery of products and services. Academic applied research has led to cumulative, incremental advances that have been of great importance to whole industries; individual companies have also benefited from university-based researchers working to solve discrete practical problems related to their businesses. University-based research centers, with industry participation, have become an important vector for transferring the results of "directed" basic research and applied research to industry.

Most of the applied research and "directed" basic research performed at universities is funded by federal agencies (e.g., the U.S. Department of Defense, U.S. Department of Energy, and the U.S. Department of Health and Human Services) looking for solutions to specific problems. Industry funds a small portion of applied research in universities, both directly and indirectly by supporting university-based research centers that are largely federally funded. Although the portion of academic research funded by industry increased in the 1990s, it is still only about 7 percent of the total.

Universities have benefited greatly from interactions with industry. Despite minimal financial support from industry, questions posed by industry often reveal gaps in knowledge that can be addressed by long-term academic research, thus stimulating fundamental research in many fields. This is apparent in high-technology industries (e.g., network systems and communications and the medical devices and equipment industries). But even in other industries (e.g., financial services or transportation, distribution, and logistics services) that perform little if any R&D but require lots of technology, industry challenges can be important stimuli to both basic and applied research.

The impact of university-based research on industry is not limited to research in the natural sciences and engineering. Research in the social sciences, broadly defined, has also made major contributions to industrial success. Research on consumer behavior, for example, has influenced industry decision-making processes in marketing, product design, and the setting of technical priorities; research in economics has informed regulatory decisions, merger and acquisition strategies, the development of financial products, and trade, monetary, and fiscal policy; research in mathematics has had a direct impact on information technology but has also been crucial to other fields (e.g., cryptography) that affect personal and business transactions daily. In these and other fields, the cumulative effects of academic research have led to changes in the legal and regulatory frameworks essential to successful innovation.

Often the contributions of academic research to industry are mediated through other disciplines or embedded in technologies, products, and services derived from other industries. The five industry studies show how academic research in physics, biology, and chemistry has led to new knowledge and capabilities in microelectronics, genetic engineering, and other fields that have directly contributed to the creation of high-value, high-technology products and services. They also show how contributions from academic research to major cross-sector technologies—information technology in particular—have directly benefited multiple industries. Information technology is critical to the technical and market performance of aircraft and medical devices, for example, and has profoundly changed the structure and performance of the financial services industry and the transportation, distribution, and logistics services industry. The industry studies also reveal the multidisciplinary character of many innovations in products and services. The development of new medical devices, for example, requires not only advances in the life sciences, but also advances in physical sciences and engineering. Many innovations in the network systems and communications industry depend on complementary progress in several fields of engineering and in the physical, social, and behavioral sciences.

The committee's review clearly indicates that academic research provides benefits to industry and has had a long-term, positive impact on industry performance. However, it is difficult to identify specific mechanisms by which this impact can be maximized for several reasons. First, the nature of university-industry interactions varies from industry to industry; each of the industries studied has a distinctive environment and poses different challenges for university researchers. Second, research competencies, ability to interface with industry, quality of infrastructure, and many, many other circumstances vary from one university to another. Third, in general, *companies*, not *industries*, interface with universities. Companies in a given industry also vary in their ability to manage that interface, in their expectations of what academic researchers can provide, in the complexity of their research needs, and in their time horizons.

All of these factors vary over time and under different circumstances (e.g., economic cycles). When this study began, high-technology industries, such as network systems and communications, were booming, attracting academic researchers and potential graduate students to well funded industry laboratories, growing operations, and new start-up companies. When the study came to a close, this same industry had suffered decreased sales, lower stock prices, lower investments, cutbacks in research funding, and lower employment. As this real-world example shows, the unique characteristics of individual academic institutions and the changes wrought by economic cycles both affect the impact of academic research.

GENERAL RECOMMENDATIONS

The general recommendations in this study address six challenges highlighted in the industry studies: (1) ensuring that universities remain repositories of expertise and resources in many disciplines by maintaining a balance of research projects; (2) cultivating interactions between academic and industry researchers; (3) harnessing academia's broad disciplinary base and potential for cross-disciplinary research and training to meet the needs of service businesses more effectively; (4) assessing the impact on the core research and educational missions of universities of an increasing emphasis on intellectual property development and management; (5) strengthening the contributions of academic research to both regulatory agencies and the overall understanding of how regulation and deregulation affect industrial growth and development; and (6) increasing the contribution of academic research to the management of information for private gain and/or public benefit in the information age.

General Recommendation 1. Because the contributions of academic research are diverse and often indirect, a broad and balanced portfolio of academic research should be maintained. Recent trends in federal funding indicate that funding levels for research in the physical sciences, engineering, and the social and behavioral sciences should be increased.

- Congress and the administration should restore the balance in federal funding of academic research by increasing support for research in the physical sciences, engineering, and the social and behavioral sciences to complement and leverage the results of recent heavy investments in the life sciences and medical sciences.
- Federal funding of academic research should continue to emphasize long-term basic research, as well as applied research (typically funded by mission agencies). Multidisciplinary research should be encouraged through support of project-specific research teams and other institutionalized mechanisms, such as engineering research centers and other university-industry research centers.

General Recommendation 2. Industries and universities should continue to explore mechanisms and pathways for bringing the benefits of academic research to industry, keeping in mind that what works well in one industry may not work well in another. Both partners should experiment with new approaches. University-industry research linkages should be adaptable, and universities should be on the lookout for opportunities to link up with new industries and explore leading-edge industry research activities and challenges.

Given the importance of personal relationships among academic and industrial researchers for productive collaboration and knowledge transfer, universities and industry should foster interactions between university- and industry-based scientists and engineers in the following ways:

- A major program of fellowships should be established to attract and support graduate students in science and engineering.
- Sabbatical programs should be established and/or expanded to encourage academic and industry researchers to spend time in each other's home research settings.
- More balanced participation by academic researchers and their industry counterparts in major conferences on specific sectors, technical systems, and disciplines should be encouraged.
- New ways of supporting personal interactions across academia-industry boundaries, including using technology to support collaboration, should be explored.
- University-industry research centers should be structured to facilitate close interaction between academic and industry researchers.
- Academic reward structures, such as promotion and tenure criteria, should be reviewed and modified (as necessary) to encourage and reward researchers who attract research support from industry and/or address significant research questions of direct importance to industry.
- Intellectual property rights policies and practices that facilitate productive research collaboration with industry should be promulgated at universities.

General Recommendation 3. The ability of academic researchers to contribute to services industries and the receptivity of leaders in the services industries to the potential contributions of academic research must both be improved. The following steps would have immediate benefits:

- Academic research contributions and capabilities relevant to each industry should be documented and promoted in the targeted communities to educate senior managers about how academic research might improve company performance in the marketplace.
- Common legal frameworks acceptable to industry and academia should be established detailing the terms of confidentiality and related conditions to facilitate academic researchers' access to operational networks and real-time data.
- Federal mission and regulatory agencies with primary responsibility for the services industries (e.g., Securities and Exchange Commission, Internal Revenue Service, Federal Communications Commission, and U.S. Department of Health and Human Services) should consider funding academic research in ways that encourage greater participation by the services industries. Engineering research centers funded by the National Science Foundation and university transportation centers funded by the U.S. Department of Transportation could serve as models.

General Recommendation 4. Individual researchers and organizations, such as the Association of University Technology Managers, that gather data on university research and technology-transfer activities should continue to monitor and assess the effectiveness of incentives for transferring academic research results (particularly intellectual property policies and practices) and the impact of entrepreneurial activity by academic researchers on the traditional university missions of education, research, and service. The following issues should be addressed:

- The costs to institutions of patenting research results, including the costs of maintaining and defending patents, should be assessed and compared to the benefits, in terms of income from licenses and royalties.
- Steps being taken to disseminate patent information to improve the chances of commercialization should be reviewed and best practices identified.
- Best practices in the long-term management of patent inventories should be shared among research institutions.
- The effectiveness of technology transfer via patented inventions should be assessed and compared to transfer via more traditional mechanisms, such as publications. The benefits to faculty and universities should also be compared.
- The impact of university-industry research collaboration and technology transfer activities on undergraduate, graduate, and continuing education, the composition of academic research, the stability of academic research funding, the private and social returns from academic research, the many traditional service roles of the university, and other related issues should be assessed.

General Recommendation 5. Government regulatory agencies, including the Food and Drug Administration, the Environmental Protection Agency, the Federal Communications Commission, and the Securities and Exchange Commission, should be encouraged to maintain and strengthen their productive interaction with academic researchers and to continue to explore new mechanisms for bringing scientific and engineering advances, including scientifically based concepts and tools, to bear more rapidly and effectively on regulatory processes.

General Recommendation 6. Government, industry, and universities should work together to meet the challenges and opportunities created by information technologies. The following steps would be beneficial:

- Boost federal funding for fundamental research in information technologies, as part of an effort to redress the imbalance in federal funding for various disciplines in academic research.

- Increase public and private sector investment in software research, with an emphasis on (1) engineering methods for assessing and improving quality and (2) software that is more flexible and responsive to changing business conditions.
- Support more interdisciplinary research on existing and potential information technologies that combines engineering methods and the social and behavioral sciences.

Biographical Information

JEROME H. GROSSMAN (chair) is senior fellow and director of the Health Care Delivery Project at Harvard University, where he brings his expertise in the health care system and information technology and his experience in community services to innovations and reforms in the medical care delivery system. He is also chairman emeritus of New England Medical Center, where he was chairman and CEO from 1979 to 1995, professor of medicine at Tufts University School of Medicine, and an honorary physician at the Massachusetts General Hospital.

Dr. Grossman is known for his leadership in the evolving role of academic medical centers in American medicine. In 1988, he founded the Health Institute at New England Medical Center, an organization involved in research and development programs and practical applications in medical outcomes, functional health status, the relationship of doctors and patients, and related areas. A member of the Institute of Medicine, he has chaired four committees studying utilization management and guidelines. In 1990, he was named a director of the Federal Reserve Bank of Boston and was appointed chair from 1994 to 1997. In 1999, he was appointed to the National Academies Council on Government-University-Industry Research Roundtable (GUIRR). In 2000, he was elected trustee of the Committee on Economic Development and is cochair of the Subcommittee on the Future of Employer-Based Health Insurance, Subcommittee on Education Policy, and Committee on the Future Supply of Scientists and Engineers. In 2001, he was invited to participate in the activities of the National Bureau of Economic Research.

ALFRED V. AHO is vice president of computer sciences research at Bell Laboratories, Lucent Technologies, Murray Hill, New Jersey. Prior to this appointment, Dr. Aho was professor and chair of the Computer Science Department at Columbia University, and before that he was general manager of the Information Sciences and Technologies Research Laboratory at Bellcore in Morristown, New Jersey. Dr. Aho received a B.A.Sc. in engineering physics from the University of Toronto and a Ph.D. in electrical engineering (computer science) from Princeton University.

Dr. Aho's research has focused on algorithms, compilers, database systems, and programming tools. He has written more than 60 research papers and published 10 books that are widely used around the world in computer science research and education. He is a coinventor of the AWK programming language and several UNIX programming tools. Dr. Aho is a member of the National Academy of Engineering, and a fellow of the American Association for the Advancement of Science, the Association for Computing Machinery, Bell Laboratories, and IEEE. He has received honorary doctorates from the University of Helsinki and the University of Waterloo for his contributions to computer science research. He has also been a distinguished lecturer at many leading universities and is an active member of a number of national and international advisory boards and committees.

CYNTHIA BARNHART is codirector of the Operations Research Center at the Massachusetts Institute of Technology (MIT), where she has been a faculty member since 1992. Her research activities have focused on the development of planning models and algorithms to improve operations at airlines, railroads, trucking firms, and intermodal partnerships. Problems she has addressed include airline crew scheduling, airline fleet assignment, less-than-truckload routing and scheduling, truckload freight routing, intermodal (e.g., rail-air-ship) freight transportation planning, express freight service design, and rail service design. Professor Barnhart has served as an associate editor for *Operations Research, Transportation Science,* and *Management Science*, as a board member for INFORMS (Institute for Operations Research and the Management Sciences), and as president of the Women in Operations Research and Management Science Forum. She has been awarded the Mitsui Faculty Development Chair, the Junior Faculty Career Award from the General Electric Foundation, and the Presidential Young Investigator Award from the National Science Foundation. Her work has been published in several books and research journals, including *Transportation Science*, *Operations Research*, *Naval Research Logistics*, *Journal of Business Logistics*, *Mathematical Programming*, *Computational Optimization and Applications,* and *Annals of Operations Research.*

ROBERT E. BIXBY is a research professor and the Noah Harding Professor Emeritus of Computational and Applied Mathematics at Rice University and technology fellow of ILOG S.A. He earned his B.S. in industrial engineering from the University of California, Berkeley (1968) and his M.S. and Ph.D. in operations research from Cornell University (1971 and 1972). His research interests include the solution of large-scale linear programming problems, the application of linear programming methods to integer programming, parallel methods for integer and linear programming, and algorithms for combinatorial optimization problems. In one real-world application, Dr. Bixby and coworkers from Princeton and Rutgers implemented a new algorithm for the solution of the linear programming relaxation of a 13-million-variable airline crew scheduling model.

In addition to his academic work, Dr. Bixby cofounded a software company in 1988 called CPLEX Optimization, Inc., that markets algorithms for linear and mixed-integer programming. CPLEX Optimization was recently acquired by ILOG S.A, of which Dr. Bixby is a director. Besides its commercial applications, the CPLEX optimizer is used by universities throughout the world in education and research in integer and linear programming. Dr. Bixby is affiliated with many scientific organizations, committees, and publications and is currently a member of the Mathematical Programming Society, the Operations Research Society of America, and the Society for Industrial and Applied Mathematics (SIAM) and chair of the Mathematical Programming Society Publications Committee. He has published more than 60 papers and technical reports.

LILLIAN C. BORRONE retired as the assistant executive director of the Port Authority of New York and New Jersey in December 2000. In that position, she advised the executive director and Board of Commissioners on various policy issues, including international trade development, real estate acquisition and disposition for maritime, aviation, and mixed-use development projects and transportation capital project management. For 12 years, Mrs. Borrone was director of the Port Commerce Department, which oversees the agency's vast marine terminals, waterfront development, and international relations responsibilities. During her tenure, the maritime activities of the Port Authority increased significantly.

Mrs. Borrone currently chairs the U.S. Department of Transportation Advisory Council to the Bureau of Transportation Statistics. She is past chair of the National Research Council (NRC) Transportation Research Board and a past member of the NRC Marine Board Executive Committee, secretary treasurer of the Board of the Eno Transportation Foundation, and a member of the National Academy of Engineering and the National Academy of Public Administration.

In 2001, Mrs. Borrone was inducted into the Maritime Association Hall of Fame. She has also been honored with the Traffic Club of New York Transportation Person of the Year 2000 Award and the Containerization Institute "Connie Award." She was a Year 2000 Executive Women of New Jersey "Salute to Policy Makers" honoree and one of New Jersey's Top Ten Business Women of 2000, selected by the *Newark Star Ledger*. Mrs. Borrone was also a recipient of the Port Authority of New York and New Jersey's Robert F. Wagner

Distinguished Service Medal. She holds an M.S. in civil engineering/ transportation management from Manhattan College and a B.A. in political science from American University.

A. RAY CHAMBERLAIN, Ph.D., P.E., is vice president of Parsons Brinckerhoff, a worldwide engineering consulting firm. He was vice president for freight policy of the American Trucking Associations, Inc., from 1994 to 1998, and executive director of the Colorado Department of Transportation from 1987 to 1994. He also served terms as president of the American Association of State Highway and Transportation Officials, chairman of the Executive Committee of the Transportation Research Board, and president of the National Association of State Universities and Land Grant Colleges. From 1969 to 1980, he was the president of Colorado State University. Dr. Chamberlain earned three degrees in engineering and has comprehensive knowledge of surface transportation issues.

JOHN M. CIOFFI received a B.S.E.E. in 1978 from the University of Illinois, Champaign, and a Ph.D.E.E. in 1984 from Stanford University. After graduation, he was a modem designer at Bell Laboratories from 1978 to 1984, and a disk read-channel researcher at IBM from 1984 to 1986. He has been on the faculty of Stanford since 1986, where he is now a tenured associate professor. He founded Amati Communications Corporation in 1991 (purchased by Texas Instruments in 1997) and was an officer and director of the company from 1991 to 1997.

Dr. Cioffi's area of interest is high-performance digital transmission. He has published more than 200 papers and holds more than 40 patents, most of which are widely licensed, including basic patents on DMT, VDSL, and V-OFDM. He was elected a member of the National Academy of Engineering in 2001 and is a fellow of the IEEE. He has received many awards: the 2001 IEEE Kobayashi Medal; the 2000 IEEE Millennium Medal; the 2000 IEEE J.J. Tomson Medal; the 1999 U. of Illinois Outstanding Alumnus; the 1995 Outstanding Achievement Award of the American National Standards Institute; and the 1991 IEEE *Communications Magazine* Best Paper Award. Dr. Cioffi was a National Science Foundation Presidential Investigator from 1987 to 1992 and has served in a number of editorial positions for IEEE magazines and conferences. He is currently on the boards or advisory boards of BigBand Networks, Coppercom, GoDigital, Ikanos, Ionospan, Ishoni, IteX, Marvell, Kestrel, Charter Ventures, and Portview Ventures. He is a member of the National Research Council Computer Science and Telecommunications Board.

PAUL CITRON is vice president of technology policy and academic relations at Medtronic, Inc. Prior to this appointment, he was vice president of science and technology, responsible for corporate-wide assessment and coordination of technology and for establishing and prioritizing corporate research. He was awarded a B.S. in electrical engineering from Drexel University in 1969 and an M.S. in electrical engineering from the University of Minnesota in 1972. Mr. Citron was elected a member of the National Academy of Engneering in 2003. He was elected founding fellow of the American Institute of Medical and Biological Engineering (AIMBE) in January 1993, has twice won the American College of Cardiology Governor's Award for Excellence, and in 1980 was inducted as a fellow of the Medtronic Bakken Society. He was voted IEEE Young Electrical Engineer of the Year in 1979. He has numerous publications and eight U.S. medical device patents to his credit. In 1980, he was given Medtronic's "Invention of Distinction" Award for his role as co-inventor of the tined pacing lead. He is a member of seven advisory boards and committees.

COLIN CROOK retired from Citicorp in July 1997 as chief technology officer. He now advises clients worldwide on issues involving information technology, business investments, the use of complex, adaptive-systems theory in business, security, and global financial enterprises. Mr. Crook is a senior fellow of the Wharton School of the University of Pennsylvania and chairman of its Information Technology Committee. He is a fellow of the Royal Academy of Engineering (elected in 1981) and a board member of several companies and nonprofit institutes. During the course of his career, Mr. Crook has worked in many parts of the world, has been involved with virtually all key information technologies, and has held a wide range of management positions.

ANTONIO L. ELIAS is executive vice president and general manager of Orbital Sciences Corporation, where he has been senior vice president for advanced programs, chief technical officer (from 1996 to 1997), and corporate senior vice president (from 1992 to 1996). As first vice president for engineering, he led the technical team that designed and built the Pegasus air-launched booster and was the launch-vehicle operator on the carrier aircraft for the rocket's first and fourth flights. He also led the design teams of Orbital's APEX and SeaStar satellites and the X-34 hypersonic research vehicle.

In 1980, Dr. Elias joined the faculty of the Massachusetts Institute of Technology (MIT), where he held the Boeing Chair in the Department of Aeronautics and Astronautics; he taught courses in control systems, spacecraft design, and computer hardware and software design, and conducted research in computer-aided engineering and air traffic control management. During the 1970s, Dr. Elias worked on the design of the Space Shuttle Orbiter avionics system at Draper Laboratory, where he originated the terminal area energy management (TAEM) guidance strategy, which is currently used for Shuttle landings.

Dr. Elias holds a B.S., M.S., E.A.A., and Ph.D. from MIT. He is a fellow of the American Institute of Aeronautics and Astronautics (AIAA), 1991 AIAA Engineer of the Year, and recipient of the AIAA Aircraft Design Award, American Astronomical Society Brouwer Award, and corecipient of the National Medal of Technology and the National Air and Space Museum Trophy. Dr. Elias was elected a member of the National Academy of Engineering in 2001. He is the author of several publications, holds several patents, and is a licensed radio amateur and commercial pilot.

DONALD M. ENGELMAN is Eugene Higgins Professor of Molecular Biophysics and Biochemistry at Yale University. His scientific interests are focused on the structure and function of proteins and biological membranes. Dr. Engelman holds an M.S. and Ph.D. in molecular biophysics from Yale. As a member of the Board of Directors of the Stryker Corporation (Kalamazoo, Michigan), he participated in a collaboration with a biotechnology firm to bring a new product, OP-1, to market, based in part on science in the academic sector. He is a member of the National Academy of Sciences, chair of the Science and Technology Steering Committee at Brookhaven National Laboratory, and a member of several National Academies panels; he also teaches biochemistry to undergraduates. He is a past chair of his department, acting dean of Yale College, and director of the Division of Biological Sciences.

DAVID J. FARBER is the Alfred Fitler Moore Professor of Telecommunication Systems at the University of Pennsylvania, where he has appointments in the Computer Science and Electrical Engineering departments. He was responsible for the design of the DCS system, one of the first operational message-based fully distributed systems and is one of the authors of the SNOBOL programming language. Dr. Farber was one of the principals in the creation and implementation of CSNet, NSFNet, BITNET II, and CREN and was instrumental in the creation of the NSF/DARPA-funded Gigabit Network Testbed Initiative; he was chair of the Gigabit Testbed Coordinating Committee. His background includes positions at Bell Laboratories, the RAND Corporation, Xerox Data Systems, University of California-Irvine, and the University of Delaware.

Dr. Farber is a fellow of IEEE and a member of the boards of directors of both the Electronic Frontier Foundation and the Internet Society. He was a 10-year alumnus of the Computer Science and Telecommunications Board (CSTB) of the National Research Council and a recipient of the 1995 SIGCOMM Award (for seminal contributions to the field of computer networks and distributed computer systems) and the 1996 John Scott Award (for contributions to humanity). He is a fellow of the Japan Glocom Institute and of the Cyberlaw Institute, and a member of the U.S. Presidential Advisory Committee on High-Performance Computing and Communications, Information Technology, and the Next-Generation Internet.

ANNETINE C. GELIJNS is director of the International Center for Health Outcomes and Innovation Research and an associate professor of surgical sciences in the Department of Surgery, College of Physicians and Surgeons, and the

School of Public Health, Columbia University. Her current research is focused on the factors driving the rate and direction of innovative activity in medicine, technological change and its relation to health-care costs, and measuring the outcomes of clinical interventions. Before coming to Columbia in 1993, she directed the Program on Technological Innovation in Medicine at the Institute of Medicine of the National Academies. From 1983 to 1987, she worked for the Steering Committee on Future Health Scenarios and for the Health Council, the Netherlands. Dr. Gelijns has been a consultant to various national and international organizations, including the World Health Organization and the Organisation for Economic Co-operation and Development. She has a Ph.D. from the medical faculty, University of Amsterdam, and bachelor's and master's degree in law from the University of Leyden, the Netherlands.

CAMILO C. GOMEZ received a B.S. in 1981 in electrical engineering and a Ph.D. in 1986 in physics from the Massachusetts Institute of Technology. From 1986 to 1991, he was principal investigator at Los Alamos National Laboratory in the Los Alamos Laser Fusion Program. From 1992 to 1995, he worked in Equity Derivatives at Lehman Brothers developing program trading systems and valuation models for exotic derivatives. In 1995, he cofounded CASA and headed the Investment Analytics Group, which focused on finance, risk management, and corporate dynamics. Recently, he has been working on the application of advanced analytical methods to the measurement and management of risk in the trade receivable space.

DANIEL S. GREGORY has been associated with the Greylock Organization, a Boston venture capital firm, since its formation in 1965, and has been a partner of the Greylock Management Corporation; from 1976 to 1991, he was managing partner of the partnerships. Mr. Gregory served in Governor William Weld's cabinet as secretary of economic affairs until January 1992; he was the first chair and is still a member of the Governor's Council for Economic Growth and Technology.

In 1983 and 1984, Mr. Gregory was president, and then chair, of the National Venture Capital Association in Washington, D.C. He is a graduate of Wesleyan University and was the recipient of Wesleyan's Distinguished Alumni Award in 1986. He served in the U.S. Navy as a deck officer from 1952 to 1955 and graduated from the Harvard Graduate School of Business Administration in 1957. Mr. Gregory has served on the boards of directors of numerous emerging, high-tech companies, including Avid Technology, Genetics Institute, New England Business Services, Teradyne Corporation, and Vertex Pharmaceuticals. He is a trustee of Wellesley College, Thompson Island Outward Bound School, Mystic Seaport Museum, and Woods Hole Oceanographic Institution.

KENNETH C. HALL received his S.B., S.M., and Sc.D. from the Department of Aeronautics and Astronautics at Massachusetts Institute of Technology. From 1987 to 1990, he was an associate research engineer at the United Technologies Research Center. In 1990, he joined the faculty of the Department of Mechanical Engineering and Materials Science at Duke University; he is currently the chair of that department. Dr. Hall's primary research is on the unsteady aerodynamics and aeroelasticity of gas turbine engines and aircraft. Other areas of research include optimization techniques, structural dynamics, and animal propulsion. He is a fellow of the American Society of Mechanical Engineers and an associate fellow of the American Institute of Aeronautics and Astronautics.

GEORGE H. HEILMEIER is chairman emeritus of Telcordia Technologies (formerly Bellcore), a leading provider of telecommunications software and professional services based on world-class research. Before joining Bellcore in 1991, Dr. Heilmeier was senior vice president and chief technical officer of Texas Instruments, Inc. He holds a B.S. in electrical engineering from the University of Pennsylvania and an M.A., M.S.E., and Ph.D. in solid-state electronics from Princeton University. He has also been awarded honorary degrees by Stevens Institute and the Israel Institute of Technology (The Technion).

Dr. Heilmeier joined RCA Laboratories in 1958, where his work with electro-optic effects in liquid crystals led to the first liquid-crystal displays for calculators, watches, computers, and instrumentation. In 1968, he was honored with the prestigious David Sarnoff Award from the IEEE and the Eta Kappa Nu Award as the Outstanding Young Electrical Engineer in the United States. In 1970, he was chosen as a White House fellow working on long-range research and development planning and technology assessment as a special assistant to the Secretary of Defense. A year later, he was appointed assistant director of Defense Research and Engineering in charge of all U.S. Department of Defense programs in electronics, computer technology, and the physical sciences. Heilmeier won confirmation in 1975 as director of the Defense Advanced Research Projects Agency (DARPA), where he initiated major efforts in stealth aircraft, space-based lasers and reconnaissance systems, infrared technology, and artificial intelligence. During his tenure at DARPA, he was twice awarded the Department of Defense Distinguished Civilian Service Medal.

Dr. Heilmeier has received numerous other awards, including the Japanese Communications and Computers Prize (1990) and three major IEEE awards. In September 1991, he was awarded the National Medal of Science by President Bush for contributions to national security and competitiveness. He received the National Academy of Engineering Founders Award in 1992 and the Eta Kappa Nu Vladimir Karapetoff Eminent Member's Award in April 1993. In 1993, he received the Industrial Research Institute Medal for outstanding accomplishment in leadership of industrial research and was named the first Technology Leader of the Year by *Industry Week* magazine. In 1996, he received the John Scott Award for Scientific Achievements from the city of Philadelphia for his pioneering work in the development of liquid-crystal displays. Dr. Heilmeier is a member of the National Academy of Engineering, Defense Science Board, and the National Security Agency Scientific Advisory Board; he serves on the boards of directors of several companies.

ROBERT J. HERMANN is senior partner at Global Technology Partners, LLC. Prior to this, he was senior vice president, science and technology, at United Technologies Corporation, where he was responsible for the development of the company's technical and scientific resources and overseeing the United Technologies Research Center. He joined the company in 1982 as vice president, systems technology, in the electronics sector and later served in a series of assignments in the Defense and Space Systems groups. For 20 years, Dr. Hermann served with the National Security Agency, where he was assigned to research and development, operations, and NATO. In 1977, he was appointed principal deputy assistant secretary of defense for communications, command, control, and intelligence. In 1979, he was named assistant secretary of the Air Force for research, development and logistics and director of the National Reconnaissance Office. He received his B.S., M.S., and Ph.D. in electrical engineering from Iowa State University. Dr. Hermann is a member of the Defense Science Board, National Academy of Engineering, and Board of Directors of the American National Standards Institute; he is chair of the Board of Directors for Draper Laboratory

ADAM B. JAFFE is Fred C. Hecht Professor of Economics and Dean of Arts and Sciences at Brandeis University. He is also coordinator of the Innovation Policy and the Economy group of the National Bureau of Economic Research (NBER). At NBER, he was also principal investigator for a National Science Foundation research project, funded through NBER, to compile a comprehensive database on patents and patent citations and use these data to document the flows of technological knowledge across time, industries, and geographic areas. Previously, he was assistant and then associate professor of economics at Harvard University. From 1992 to 1994, he was visiting professor at the John F. Kennedy School of Government at Harvard.

Professor Jaffe's areas of specialization are the economics of technological change, the economic impact of universities, and the economics of regulated industries. He has published papers on industrial research and development, the economics of basic research and universities, incentive regulation and regulatory reform, and the determinants of the diffusion of new technologies. In 1990–1991, Dr. Jaffe took leave from Harvard to serve as senior staff economist to the President's Council of Economic Advisers in Washington, D.C., where he was responsible for energy policy, technology policy, and regulatory policy. He received his S.B. in chemistry (1976) and his S.M. in technology and policy (1978) from Massachusetts Institute of Technology, and his Ph.D. in economics from Harvard University (1985).

JACK L. KERREBROCK is professor emeritus of aeronautics and astronautics at Massachusetts Institute of Technology (MIT). He joined the faculty of MIT in 1960, where he remained as professor, department head, and associate dean (except for two years as associate administrator for aeronautics and space technology at National Aeronautics and Space Administration [NASA]). Dr. Kerrebrock is a member of the National Academy of Engineering and was chair of the National Research Council (NRC) Committee on the Space Station. He has also served as member and chair of numerous other NRC and NASA committees. Dr. Kerrebrock received his Ph.D. from the California Institute of Technology.

KENT KRESA was elected president of Northrop Grumman in 1987, chief executive officer in January 1990, and chairman in September 1990. Before joining Northrop Grumman in 1975, he worked at the Defense Advanced Research Projects Agency, where he was responsible for applied research and development programs in the tactical and strategic defense arena. From 1961 to 1968, he was associated with the Lincoln Laboratory at Massachusetts Institute of Technology (MIT), where he worked on ballistic missile defense research and reentry technology.

Mr. Kresa has received many industry and government honors, most recently, the Private Sector Council's 2001 Leadership Award (for commitment to improve governmental efficiency), the Aerospace Historical Society International von Kármán Wings Award (for contributions to the industry), and (with Northrop Grumman) "Manufacturer of the Century" by the California Manufacturers and Technology Association. Mr. Kresa is serving a one-year term as president of the American Institute of Aeronautics and Astronautics (AIAA). He is a member of the National Academy of Engineering, the past chair of the Board of Governors of the Aerospace Industries Association, and chair of the Defense Policy Advisory Committee on Trade. Mr. Kresa received a B.S., M.S., and E.A.A., all in aeronautics and astronautics, from MIT.

BLAKE D. LEBARON, who earned a Ph.D. in economics from the University of Chicago, is professor of finance at the Graduate School of International Economics and Finance of Brandeis University, a faculty research fellow at the National Bureau of Economic Research, a member of the external faculty of the Santa Fe Institute, and a Sloan fellow. Before moving to Brandeis, Dr. LeBaron was associate professor of economics at the University of Wisconsin-Madison. In 1993, he was director of the Economics Program at the Santa Fe Institute. Dr. LeBaron's research has been concentrated on the nonlinear behavior of financial and macroeconomic time series. His current interests are the quantitative dynamics of interacting systems of adaptive agents and how these systems replicate observed real-world phenomena and the observed behavioral characteristics of traders in financial markets, including strategies and policy questions. In general, he seeks to discover the empirical implications of learning and adaptation as applied to finance and macroeconomics.

JOHN M. MULVEY is a professor in the Department of Operations Research and Financial Engineering and a founding member of the Bendheim Center for Finance at Princeton University. His specialty is the application of large-scale optimization models and algorithms, with an emphasis on strategic financial planning. He has implemented integrated risk-management systems for many companies, linking the key risks to the organization and assisting in high-level decisions. In addition, he has designed a number of significant planning systems for government agencies, including the Office of Tax Analysis for the Treasury Department, the Joint Chiefs of Staff in the U.S. Department of Defense, and personnel planning for the U.S. Army. He holds a B.S. (1969) in general engineering and an M.S. (1969) in computer science from the University of Illinois, Urbana, and an M.S. (1974) and Ph.D. (1975) in management science from the University of California, Los Angeles. He has edited three books and published over 100 scholarly papers.

HOMER PIEN is chief executive officer of SRU Biosystems, Woburn, Massachusetts; previously, he was chief technology officer of Medical OnLine, Lexington, Massachusetts. Dr. Pien also spent 10 years at Draper Laboratory, both as manager of biomedical technologies and head of the Image Recognition Systems Laboratory. Concurrently, he was director of technology at the Center for Innovative Minimally Invasive Therapy, a research consortium comprising Massachusetts General Hospital, Brigham and Women's Hospital, Massachusetts Institute of Technology (MIT), and Draper Laboratory. Prior to joining Draper, he was with MIT Lincoln Laboratory for five years. Dr. Pien received his B.S. in mathematics from the University of Illinois, his M.S. and Ph.D. in computer science from Northeastern University, and an M.S. (Management) from the MIT Sloan School. He is an adjunct professor in the Graduate College of Computer Science at Northeastern University and a member of the Department of Radiology at Massachusetts General Hospital.

H. DONALD RATLIFF is UPS and Regents Professor of Industrial and Systems Engineering, as well as executive director of the Logistics Institute, at Georgia Institute of Technology. He is also cofounder and president of CAPS Logistics, a leading-edge supplier of logistics software. He has published more than 50 refereed papers, designed a variety of computer software applications addressing logistics issues, delivered more than 200 invited presentations at national and international meetings, and has been a consultant for more than 40 companies on a wide range of issues related to the movement and storage of products. He has supervised 20 Ph.D. students, 11 of whom currently hold faculty appointments at various universities. Dr. Ratliff has been editor in chief of the *Journal of Operations Research*, area editor for *Optimization*, departmental editor for applied optimization in *IIE Transactions*, and associate editor of *Management Science*. He was awarded the 1991 Outstanding Research Award of the Institute of Industrial Engineers for his work in logistics and is a fellow of the Institute of Industrial Engineers and a member of the National Academy of Engineering.

J. DAVID ROESSNER is professor of public policy at Georgia Institute of Technology, codirector of the Technology Policy and Assessment Center at Georgia Tech and, since September 1995, program manager for technology policy at SRI International. Prior to joining the Georgia Tech faculty in 1980, he was principal scientist and group manager for industrial policy and planning at the Solar Energy Research Institute in Golden, Colorado; policy analyst with the National Science Foundation (NSF) R&D Assessment Program; acting leader of the Working Group on Innovation Processes and Their Management in the Division of Policy Research and Analysis at NSF; and research associate at the Bureau of Social Science Research, Inc. He began his professional career as a development engineer for Hewlett-Packard in Palo Alto, California.

Dr. Roessner received his B.S. from Brown University, his M.S. in electrical engineering from Stanford University, an M.S. and his Ph.D. in science, technology, and public policy from Case Western Reserve University. His research interests include national technology policy, the evaluation of research programs, the management of innovation in industry, technology transfer, and indicators of scientific and technological development. In addition to numerous technical reports, he has published papers in many policy-oriented journals. From 1987 to 1995, he was a U.S. editor of *Research Policy* and is still an advisory editor for that journal. He is also a member of the editorial boards of the *Journal of Technology Transfer* and *Technology Studies*; principal author *of The Impact of Office Automation on Clerical Employment, 1985–2000* (Quorum Books, 1985); editor of *Government Innovation Policy: Design, Implementation, Evaluation* (St. Martin's Press, 1988); editor of a special issue of *Research Policy*, published in 1989, devoted to the evaluation of government innovation programs; and co-editor of a 1996 special issue of *Research Policy*, "Evaluation of Industrial Modernization."

Dr. Roessner is a member of the Committee on Science, Engineering and Public Policy of the American Association for the Advancement of Science (AAAS) and, in 1996, was elected a fellow of the AAAS. He has served as a consultant to the Congressional Office of Technology Assessment; the U.S. Departments of Defense, Energy, and Commerce; National Science Foundation; Jet Propulsion Laboratory; Oak Ridge National Laboratory; Sandia National

Laboratories; Los Alamos National Laboratory; the General Accounting Office; the RAND Corporation; and SRI International.

ROBERT F. SPROULL, vice president and fellow at Sun Microsystems Laboratories, leads a section of the laboratory in Burlington, Massachusetts. Since his undergraduate days, he has been building hardware and software for computer graphics, including clipping hardware, an early device-independent graphics package, page description languages, laser printing hardware and software, and window systems. He has also been involved in VLSI design, especially of asynchronous circuits and systems. Before joining Sun, he was a principal with Sutherland, Sproull & Associates, an associate professor at Carnegie Mellon University, and a member of the Xerox Palo Alto Research Center. He is co-author, with William Newman, of *Principles of Interactive Computer Graphics* (McGraw-Hill, 1981) and an author of *Logical Effort* (Morgan Kaufmann Publishers, 1999). Dr. Sproull is a member of the National Academy of Engineering, has served on the U.S. Air Force Scientific Advisory Board, is a special partner of Advanced Technology Ventures, and is a member of the Board of Directors of Alphatech, Inc.

MORRIS TANENBAUM retired as vice chairman of the board and chief financial officer of AT&T on June 30, 1991. He began his career there in 1952 as a member of the technical staff of Bell Telephone Laboratories, was promoted to department head in the Research Division in 1955, to director of the Solid State Device Laboratory in 1962, director of research and development for the Western Electric Company in 1964, followed by vice president of the Engineering Division and vice president of manufacturing, before returning to Bell Labs in 1975 as executive vice president responsible for the development of customer services, switching and transmission equipment, and electronics technology. Dr. Tanenbaum holds seven patents and has contributed to numerous books and technical journals. He pioneered the use of silicon as a commercial semiconductor material with the invention of the diffused-base silicon transistor, which became the principal building block of integrated circuitry. Later he supervised the group that discovered the first practical superconducting materials for high field strength magnets.

Dr. Tanenbaum has served on many corporate boards and is a director of the New Jersey Performing Arts Center, a member of the Corporation of the Massachusetts Institute of Technology, an associate trustee of Battelle Memorial Institute, an honorary trustee of the Brookings Institution, and trustee emeritus of the Johns Hopkins University and Tufts University. Formerly, Dr. Tanenbaum served as vice president and councillor of the National Academy of Engineering and as a member of the Governing Board of the National Research Council. He is a fellow of the American Academy of Arts and Sciences, the IEEE, the American Physical Society, the American Association for the Advancement of Science, and a member of the American Chemical Society and the Metallurgical Society of AIME. Dr. Tanenbaum earned a B.A. in chemistry from Johns Hopkins University and a Ph.D. in physical chemistry from Princeton University.

FREDERICK W. TELLING is a corporate vice president of Pfizer, Inc., and head of Corporate Policy and Strategic Management; he began his career with Pfizer in 1977. Dr. Telling received his B.A. from Hamilton College and his Master of Industrial and Labor Relations and Ph.D. in economics and public policy from Cornell University. He is vice chairman for the American Foundation for Pharmaceutical Education and an invited faculty lecturer at the Harvard University School of Public Health and Cornell University. He is involved in the work of the Council on Competitiveness, the National Association of Manufacturers, and the Institute of Medicine (IOM), where he was a contributing author to the IOM series on *Technology Innovation in Medicine*. He is a member of the Board of the Alliance for Aging Research, Bergen County United Way and Bergen County United Way Community Foundation, Biotechnology Industry Organization (BIO), California Health Care Institute, Children of Bellevue, Committee for Economic Development, HealthCare Institute of New Jersey, National Council on Aging, New York Academy of Medicine, New York Academy of Sciences, and many other organizations. He is also a trustee and a member of the Executive Committee of the New Jersey Aviation Hall of Fame; he is actively involved in aviation policy and history.

STEPHEN WOLFF is executive director of the Advanced Internet Initiatives Division of Cisco Systems, Inc., where he is responsible for seeking out, initiating, and leading Cisco's participation in partnerships at the forefront of Internet development and deployment worldwide, including U.S. Next Generation Internet Program and the Internet2 Project. Prior to joining Cisco in 1995, Mr. Wolff was director of the Division of Networking and Communications Research and Infrastructure at the National Science Foundation (NSF), where he was responsible for the National Research and Education Network (NREN) and NSFNET programs and for NSF's support of basic research in networking and communications. He was educated at Swarthmore College, Princeton University, and Imperial College; he is a member of the Association for Computing Machinery and a life member of the IEEE.